SOLID PHASE MICROEXTRACTION

SOLID PHASE MICROEXTRACTION

A PRACTICAL GUIDE

edited by
SUE ANN SCHEPPERS WERCINSKI

Varian Chromatography Systems
Walnut Creek, California

Taylor & Francis

Taylor & Francis Group
Boca Raton   London   New York

CRC is an imprint of the Taylor & Francis Group,
an informa business

Published in 1999 by
CRC Press
Taylor & Francis Group
6000 Broken Sound Parkway NW, Suite 300
Boca Raton, FL 33487-2742

International Standard Book Number-10: 0-8247-7058-7 (Hardcover)
International Standard Book Number-13: 978-0-8247-7058-7 (Hardcover)

Library of Congress Cataloging-in-Publication Data

Catalog record is available from the Library of Congress

Taylor & Francis Group
is the Academic Division of Informa plc.

Visit the Taylor & Francis Web site at
http://www.taylorandfrancis.com

and the CRC Press Web site at
http://www.crcpress.com

Preface

Analytical labs worldwide demand high sample throughput, fast reporting of results, reduced operating costs, and instruments that occupy minimum benchspace. Surprisingly, many chromatography analyses consume less time than preparing the sample for analysis. In fact, two-thirds of analysis time is typically spent on the sampling and sample preparation steps because most procedures are based on nineteenth century technologies that are time- and labor-intensive, contain multiple steps that can lose analytes, and use toxic organic solvents. To illustrate, a recent survey of HPLC and GC users stated that 90% of the respondents use two or more preparation techniques per sample [1]. As a result, integrating several sample preparation and separation methods is very difficult without some kind of human intervention. And, of course, the possibility for error (human, systematic, or contamination) occurs with each additional step.

The goal of sample preparation is to produce samples with the highest analyte concentration possible and the lowest level of contamination, thereby maximizing the analyte signal while minimizing interferences in the subsequent analysis. Obviously, this goal should be achieved with the easiest to reproduce and least costly procedure. Solid Phase Microextraction (SPME), extracts the analytes of interest without additional sampling or technician time, and consequently minimizes the chance for human or systematic error. Moreover, SPME does not require additional solvents or benchspace. It can be used for field sampling, such as streams, air, and fire residue, or it can be easily automated using a single fiber to sequentially sample from numerous vials, then desorb the sample into a gas or liquid chromatograph. It fulfills the laboratory requirements of productivity and reduced costs; therefore, it is an attractive sample preparation technique to replace the traditional techniques of static headspace, purge and trap, liquid/liquid extraction, and Sohxlet extraction.

SPME has been commercially available for only five years and new applications are being developed and published rapidly. Nevertheless, as with any rapidly developing technique, chemists have had to conduct their own literature search to determine if the technique may apply to their work. This comprehensive reference will assist readers in determining whether SPME can replace their current sample handling techniques. It is the first text to provide a full spectrum of proven SPME methods for laboratories performing routine analyses from authoritative research and methods development chemists. Readers will benefit from understanding the technique and comparing it to traditional methods for use in their own laboratories. In addition, proven SPME methods and practical tips for developing new methods will directly assist the chemist, thereby saving methods development cost and time.

This book covers three areas. The first chapters present SPME theory, a methodical approach to developing new SPME methods, and a thorough description of available fibers and the classes of compounds to which they apply. Second, specific application chapters on pharmaceutical, environmental, foods and flavors, and forensic and toxicology methods provide in-depth discussions of Solid Phase Microextraction used in these disciplines. These discussions include how SPME meets the day-to-day challenges that the analytical chemist faces and the regulatory agencies' requirements. For example, environmental laboratories must achieve the low minimum detectable quantities that are outlined in EPA or other environmental regulatory agencies' methods; moreover, laboratories require fast sample turnaround to remain competitive. For SPME to meet their needs, it must provide equivalent or better analysis time and results than their current methods. Therefore, each chapter illustrates how SPME meets specific industry requirements for individual applications. Third, Professor Janusz Pawliszyn, the inventor of the technique, and a member of his research team describe new developments in the technology and recommendations for new applications. Professor Pawliszyn and his researchers have shown vision in numerous analytical disciplines and continue to innovate new technologies for sample preparation and analyses.

The contributing authors are research scientists who initially developed the technique and industry scientists who have further developed practical methods. The contributors have pioneered the use of SPME in their specific application fields; furthermore, they frequently present and publish their work with SPME. This combination of university researchers, manufacturers' research chemists, and industry scientists balances the book's content between theory and practical applications.

Unlike much of the published material on the technique, this book emphasizes practical applications and methods development using commercially available fibers. The contributors have provided valid and reproducible applications in their area of expertise, explained SPME versus other techniques for the applications, and documented their work with supporting chromatograms, graphs, tables, and references. This book is intended for analytical chemists and laboratory managers who are looking for faster and less expensive methods to perform their analyses. Moreover, it assumes that readers are already knowledgeable in sample handling techniques used in chromatography. Potential readers include chemists who have not tried SPME and want to determine if it will work for their applications. Additionally, experienced users will value this comprehensive reference on this relatively new technique. Furthermore, the book can be used as a graduate level textbook in a sample preparation course.

I sincerely thank all contributors for their outstanding research and dedication to this project. Their efforts have stimulated the scientific community to rethink how sample preparation is performed. I would like to especially remember Mr. Brian MacGillivray, author of the environmental applications chapter, who passed away last year. Brian's enthusiasm for this project was vital in making it a reality, and his talents will be greatly missed. I thank Dr. Stephen Scypinski, co-author of the pharmaceutical applications chapter, who provided tremendous help early in my career while helping me to maintain my sense of humor. Steve was also instrumental in initially designing the book's contents. I am grateful to my colleagues at Varian Chromatography Systems, who provided a stimulating environment to create this book. I am also indebted to Professor Barry Eckhouse of St. Mary's College in Moraga, CA, who renewed my fondness for writing. Most of all, I thank my husband, Peter Wercinski, and my parents, Charles and Rosemary Scheppers, for their encouragement, patience, and love. It is to them that I dedicate this book.

Sue Ann Scheppers Wercinski

1. Ronald E. Majors. Trends in Sample Preparation. LC-GC Magazine, Vol. 14, No. 9, 1996, pp. 754-766.

Contents

Contributors

José R. Almirall, Ph.D Assistant Professor, Department of Chemistry, Florida International University, Miami, Florida

Ralf Eisert, Ph.D. Professor, Department of Chemistry, University of Waterloo, Waterloo, Ontario, Canada

Kenneth G. Furton, Ph.D. Associate Professor, Chairperson, Department of Chemistry, Florida International University, Miami, Florida

Brian MacGillivray, M.S.[†] Water Technology International Corporation, Burlington, Ontario, Canada

Janusz Pawliszyn, Ph.D. Professor, Department of Chemistry, University of Waterloo, Waterloo, Ontario, Canada

Zelda Penton, Ph.D. Senior Chemist, Varian Chromatography Systems, Walnut Creek, California

Terry L. Peppard, Ph.D.[*] Director of Flavor Analytical Services, Research and Development, Givaudan Roure Corporation, Clifton, New Jersey

Sue Ann Scheppers Wercinski, M.B.A. Varian Chromatography Systems, Walnut Creek, California

---

[†]Deceased.
[*]*Current affiliation*: Director of Regulatory and Scientific Affairs, Research & Development, Robertet Flavors, Inc., Piscataway, New Jersey

Stephen Scypinski, Ph.D   Director, Analytical Development, The R.W. Johnson Pharmaceutical Research Institute, Raritan, New Jersey

Robert E. Shirey, M.S.   Senior Research Chemist, Sample Handling R&D, Supelco, Bellefonte, Pennsylvania

Ann-Marie Smith, B.S.   Senior Scientist, Pharmaceutical and Analytical R&D, Hoffman-LaRoche, Inc., Nutley, New Jersey

Xiaogen Yang, Dipl.-Chem., Dr.rer.nat.   Group Leader, Flavor R&D, Givaudan Roure Corporation, Cincinnati, Ohio

1

Solid Phase Microextraction Theory

Sue Ann Scheppers Wercinski
Varian Chromatography Systems, Walnut Creek, California

Janusz Pawliszyn
University of Waterloo, Waterloo, Ontario, Canada

INTRODUCTION

An understanding of Solid Phase Microextraction (SPME) theory provides not only insight into the technique, but more importantly, assistance when developing and optimizing methods. First, this chapter describes the SPME components and sampling procedure to explain how the technique is performed. Second, SPME theory is highlighted to support how sensitivity is achieved and how the extraction and desorption times are determined for both direct and headspace sampling. The theory is applied for general guidelines on increasing sensitivity and decreasing extraction times. In addition, specific techniques are described in the subsequent chapters on methods development, fiber selection, and applications. Finally, great care has been taken to emphasize practical aspects of the theory throughout the chapter for realistic laboratory conditions and applications. This should assist an individual's understanding of how the technique applies to their own specific applications.

SPME COMPONENTS AND SAMPLING PROCEDURE

Solid Phase Microextraction (SPME) utilizes a short, thin, solid rod of fused silica (typically one cm long and 0.11 mm outer diameter), coated with an absorb-

1

ent polymer. The fiber is the same type of chemically inert fused silica used to make capillary GC columns and it is very stable even at high temperatures. The coated fused silica (SPME fiber) is attached to a metal rod, and both are protected by a metal sheath that covers the fiber when it is not in use. For convenience, this assembly is placed in a fiber holder and, together, the system resembles a modified syringe (Figure 1.1).

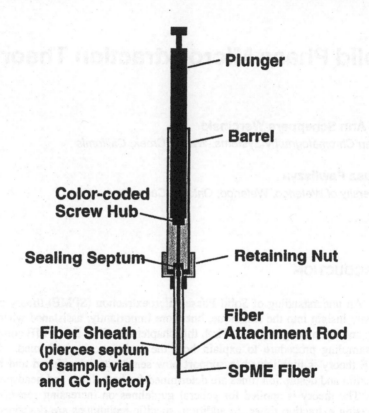

Figure 1.1 SPME fiber with holder. (Courtesy Z. Penton, Varian Associates, Inc.)

The SPME extraction technique consists of two processes: (1) analytes partition between the sample and the fiber coating, and (2) the concentrated analytes desorb from the coated fiber to an analytical instrument. To perform the extraction, an aqueous sample containing organic analytes or a solid sample containing volatile organic analytes is placed in a vial, then closed with a cap and septum. To sample, the SPME protective sheath pierces the septum then the plunger is lowered to either immerse the fiber directly into the aqueous sample or expose it to the sample headspace (Figure 1.2). The target analytes are subsequently extracted from the sample matrix into the fiber coating. After a pre-determined absorption time, the fiber is withdrawn back into the protective sheath, then the sheath is pulled out of the sampling vial. The sheath is immediately inserted in the GC or HPLC injector and the plunger is again lowered to expose the fiber. This time, the fiber is exposed to a high temperature in the injector liner (GC) where the concentrated analytes are thermally desorbed and, consequently refocused onto the GC column (Figure 1.3). In HPLC, solvents are used to desorb the analytes from the fiber. Afterwards, the fiber is withdrawn into the protective sheath and it is removed from the injector.

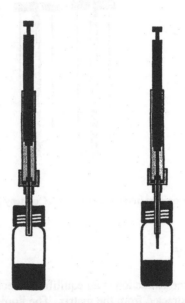

Figure 1.2 SPME headspace sampling. (Courtesy Z. Penton, Varian Associates, Inc.)

Different types of sorbents will extract different groups of analytes; therefore, many different fiber coatings have been developed. Similar to selecting an analytical GC column where "like dissolves like," a fiber is chosen based on its selectivity for certain target analytes and their volatility ranges. Nonpolar coatings (e.g., poly(dimethylsiloxane)) retain hydrocarbons very well. In contrast, polar fiber coatings (e.g., polyacrylate and carbowax) extract polar compounds such as phenols and carboxylic acid very effectively. The affinity of the fiber coating for target analytes is crucial in SPME sampling because both the matrix and fiber coating are competing for analytes. For example, a polar coating chosen to extract polar compounds from water must have a stronger affinity for the analytes than water in order for them to be extracted. The process of choosing a fiber coating for specific analyses is discussed in Chapter 3.

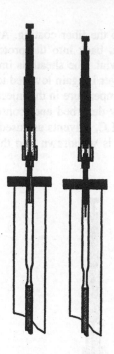

Figure 1.3 SPME desorption into a GC injector. (Courtesy Z. Penton, Varian Associates, Inc.)

SPME SENSITIVITY

Solid Phase Microextraction is an equilibrium technique; therefore, analytes are not completely extracted from the matrix. The liquid polymeric fiber coatings (e.g., poly(dimethylsiloxane)) provide a non-exhaustive liquid-liquid extraction with the convenience of the "organic phase" being attached to the fiber [1]. When a sample is placed in a closed vial, an equilibrium forms between three phases: (1) the fiber coating to the aqueous phase, (2) the headspace to aqueous phase, and (3) the fiber coating to headspace (Figure 1.4). The analyte recovery expected from SPME is related to the overall equilibrium of the three phases present in the sampling vial.

Of course, the total amount of analyte does not change during the extraction. Moreover, the distribution among the three phases after equilibrium is represented as follows:

$$C_o V_s = C_h^\infty V_h + C_s^\infty V_s + C_f^\infty V_f \tag{1.1}$$

Where: C_0 is the initial concentration of the analyte in the aqueous solution; C_h^∞, C_s^∞, and C_f^∞ are the equilibrium concentrations of the analyte in the headspace, aqueous solution, and fiber coating, respectively; and V_h, V_s, and V_f are the volumes of the headspace, aqueous solution, and fiber coating, respectively [2]. If no headspace exists in the closed vial, then $C_h^\infty V_h$, the headspace term, is omitted and the equilibrium is formed only between the aqueous solution and fiber.

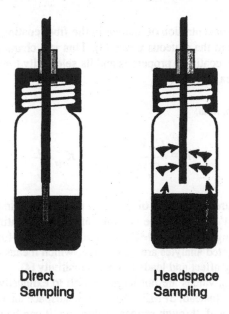

Direct Sampling   **Headspace Sampling**

Figure 1.4 SPME is a three-phase system between the solid or aqueous solution, $(C_s^\infty V_s)$, the headspace above the liquid or solid, $(C_h^\infty V_h)$, and the fiber coating, $(C_f^\infty V)$. Direct sampling (left) and headspace sampling (right) are illustrated. The partition coefficients between the three phases are $K_{fh} = \dfrac{C_f}{C_h}$, $K_{hs} = \dfrac{C_h}{C_s}$, and $K_{fs} = \dfrac{C_f}{C_s}$. (Courtesy Z. Penton, Varian Associates, Inc.)

In this section, the theory developed for fibers with liquid polymeric coatings is discussed. This rationale is also applicable to the more recently developed fibers, e.g., porous solid materials. First, the principles of direct, liquid sampling, i.e., immersing the fiber directly into the aqueous sample, will be discussed. Then, the principles of headspace sampling will be presented.

Direct Liquid Sampling

The partitioning between the fiber coating (stationary phase) and the aqueous phase is described by the distribution constant, K_{fs}:

$$K_{fs} = \frac{C_f}{C_s}$$ (1.2)

Where: C_f is the concentration of analyte in the fiber coating and C_s is the concentration of analyte in the aqueous phase [3]. This is a characteristic parameter that describes the fiber coating's properties and its selectivity toward a specific analyte, versus other matrix components.

The partition ratio, k' is:

$$k' = \frac{C_f V_f}{C_s V_s} = \frac{n_f}{n_s} = K_{fs} \frac{V_f}{V_s}$$ (1.3)

Where: n_f and n_s are the number of moles in the fiber coating and aqueous phases, respectively, and V_f and V_s are the volumes of the fiber coating and water solution. Because the coatings used in SPME have strong affinities for organic compounds, K_{fs} values for targeted analytes are quite large, which means that SPME has a very high concentrating affect and leads to good sensitivity [4].

However, K_{fs} values are not large enough to exhaustively extract most analytes in the matrix. Instead, SPME, like static headspace analysis, is an equilibrium sampling method and, through proper calibration, it can be used to accurately determine the concentration of target analytes in a sample matrix. Two different equations are used to determine the amount absorbed by the fiber, depending on the sample volume. For large sample volumes (>5 mL), the amount of analyte absorbed by the fiber coating at equilibrium is directly proportional to the initial aqueous concentration, C_o. The following equation is used when the volume of the aqueous sample, V_s, is much larger that the stationary phase volume; that is, its volume is relatively infinite to the fiber volume ($V_s >> K_{fs} V_f$).

$$n_f = K_{fs} V_f C_o$$ (1.4)

Where: n_f is the amount extracted by the fiber coating. Here, the sample volume does not need to be known (it is relatively infinite), which is ideal for field sampling and simplifies laboratory operations.

On the other hand, when sampling from a finite sample volume, such as 2–5 mL, the sample can be significantly depleted, and the amount absorbed becomes:

$$n_f = \frac{K_{fs} V_f V_s C_o}{K_{fs} V_f + V_s}$$ (1.5)

As in the case with an infinitely aqueous volume, the amount of analyte absorbed by the fiber coating is directly proportional to the initial analyte concentration, C_o. However, the additional term of $K_{fs}V_f$ is now present in the denominator. This new term decreases the amount of analyte absorbed by the fiber coating, but it is significant only when compared to V_s, the aqueous volume. In reality, this occurs only for a large distribution constant, K_{fs}, because the volume of the fiber stationary phase, V_f, is very small. If the distribution constant is very large, then $K_{fs}V_f$ >>V_s; consequently, most of the analyte will be transferred to the fiber coating [1].

Headspace Sampling

Gaseous samples and relatively clean water samples are amenable to placing the fiber directly into the sample to extract organic compounds because the analytes have a higher affinity for the fiber and interfering contaminants don't exist. On the other hand, direct sampling may not work well when sampling analytes from wastewater samples with grease or oil, or more complex samples that contain solid or high molecular weight materials such as soil or sludge. Sampling from the headspace above the sample is necessary in these cases.

As mentioned earlier, SPME equilibrium is a three-phase system that includes the solid or aqueous sample, the headspace above the liquid, and the fiber coating. The equations governing the equilibrium process between the three phases are:

$K_{fh} = \dfrac{C_f}{C_h}$ Where: K_{fh} is the partition coefficient of an analyte between the fiber coating and headspace phases, and C_f and C_h are the concentrations of the analyte in these phases.

$K_{hs} = \dfrac{C_h}{C_s}$ K_{hs} is the partition coefficient of an analyte between the headspace and aqueous phases, and C_h and C_s are the concentrations of the analyte in these phases.

$K_{fs} = \dfrac{C_f}{C_s}$ K_{fs} is the partition coefficient of an analyte between the fiber coating and aqueous phases, and C_f and C_s are the concentrations of the analyte in these phases.

As a result, the amount of analyte absorbed by the fiber coating in headspace sampling can be expressed as:

$$n_f = \frac{K_{fs} V_f V_s C_o}{K_{fs} V_f + K_{hs} V_h + V_s}$$ (1.6)

The three terms in the denominator represent the analyte capacity for each phase: fiber $(K_{fs}V_f)$, headspace $(K_{hs}V_h)$, and the sample itself (V_s). If the sample vial is completely filled with the aqueous sample; i.e., no headspace exists, then the term $K_{hs}V_h$ in the denominator can be eliminated. Consequently, the equation becomes exactly as equation (1.5) for direct sampling from a finite volume.

As expected from the equilibrium conditions, equation (1.6) states that the amount of analyte extracted by the fiber is independent of where the fiber is located in the system during the absorption step. Therefore, the amount extracted and the detection limits will be similar regardless of whether the fiber is placed in the headspace or directly in the sample, as long as the volumes of the fiber coating, headspace, and sample are kept constant [2,5]. Rather, the sampling method is chosen depending on whether or not the sample matrix contains contaminants that may interfere with the extraction, and consequently, the chromatographic separation.

Improving Sensitivity

Referring to equations (1.4) through (1.6), several parameters can be modified to increase the technique's sensitivity. One obvious approach is to increase the fiber coating thickness, thus increasing V_f. As equation (1.4) indicates for large volume samples, doubling the fiber coating can double the mass of analyte absorbed at equilibrium. Figure 1.5 clearly demonstrates that the coating thickness changes the extraction amount [5]. On the other hand, the thicker coating also increases the equilibration time simply because more analytes penetrate into a greater fiber volume. Naturally, the thinnest fiber thickness should be used to obtain the sensitivity desired in the minimum extraction time. Another option to increase the fiber coating volume is to increase the fiber length. Fibers that are 2 cm long, instead of the standard 1 cm, can be used to increase sensitivity. However, several precautions should be taken when using this fiber. These are described in Chapter 3.

Another method to increase sensitivity is to increase K_{fs}, the coating/water distribution constant of the analyte, by changing to a fiber coating that is more selective for the target analytes. The selectivity of the fiber coating for specific analytes can be improved by modifying the chemical structure of the polymer, just as using a more selective GC column phase retains specific analytes to improve the chromatographic separation. Thus, if a more polar polymer is coated onto the fiber, then more polar compounds can be analyzed at lower concentrations. Figure 1.6 illustrates this concept by showing the extraction profile of three BTEX components, each with a different distribution constant for a 56 μm thick PDMS coating. P-xylene, with the highest K_{fs}, yields the greatest response because it has a much greater affinity for the fiber coating. Consequently, analytes with a high affinity for the fiber, i.e., high K_{fs} values, do not require the use of thick fiber coatings to achieve good sensitivities. Again, using the thinnest coating possible minimizes the extraction time.

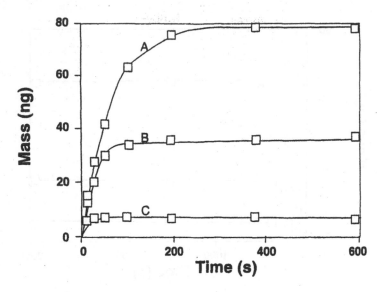

Figure 1.5 Effect of coating thickness on the absorption versus time profile for the extraction of 0.1 ppm benzene from a 2,500 rpm stirred aqueous solution. Parameters as given for Figures 1.12 and 1.14 except for fiber coating inner and outer radius; (A) 100 μm thick coating, a = 0.0055 cm, b = 0.0155 cm; (B) 56 μm thick coating, a = 0.007 cm, b = 0.0126 cm; (C) 15 μm thick coating, a = 0.0055 cm, b = 0.0070 cm. (From Ref. 5)

Finally, optimizing sample temperature also plays a key role in the method's sensitivity by changing the distribution constant, K_{fs}. Generally, increasing the sample temperature will increase the sensitivity for the higher boiling components, but decrease the sensitivity for the lower boiling components. Increasing the sample temperature creates increased diffusion coefficients and decreased distribution constants, which both lead to a faster equilibration time [5]. In turn, this effect can be applied to optimize the extraction time for a given fiber coating. In headspace SPME, an increase in the sample temperature increases the analyte concentration in the headspace, thereby providing faster extractions. Additionally, the higher temperature increases the extraction amount for high molecular weight compounds (Figure 1.7). Note that at each temperature increase, additional higher boiling components are extracted from the sample. However, the temperature increase has an adverse effect for more volatile components. The sensitivity for these is decreased due to the decreased K_{fs} at the higher temperature. Furthermore, Figure 1.8 compares the mass extracted from conventional sample heating to microwave heating. Rapid microwave heating releases analytes from the matrix quicker, yielding greater sensitivity for most of the components. Other techniques to increase sensitivity, such as salting-out, will be discussed in Chapter 2.

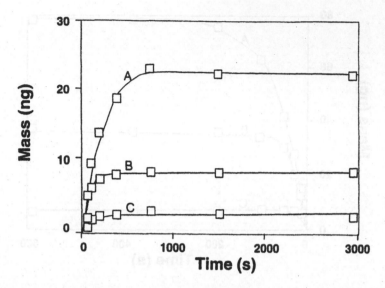

Figure 1.6 Effect of distribution constant on the absorption profile of 2,500 rpm stirred 0.1 ppm analyte extracted with a 56 μm thick coating on a 1 cm long fiber. Parameters as given for Figures 1.12 and 1.14, except for distribution constants. The curves represent: (A) *p*-xylene, $K_{fs} = 831$; (B) toluene, $K_{fs} = 294$; and (C) benzene, $K_{fs} = 125$. (From Ref. 5)

EXTRACTION SPEED FOR DIRECT AND HEADSPACE SAMPLING

Speed of Extraction for Direct Sampling

The speed of extraction is determined by how long it takes the analytes to move from the sample matrix to the fiber. This process involves: (1) the rate the analytes desorb from a solid surface if particulate matter is present, (2) the analytes migrating through the air or liquid sample, and (3) the analytes diffusing into the fiber coating [1]. One or several of these processes becomes the rate-limiting step, depending on the sampling conditions. The extraction time for direct sampling, i.e., immersing the fiber into the liquid, employing both dynamic and static conditions will be examined here.

Direct Sampling from an Agitated Aqueous Solution

Analytes in solution are extracted faster using some form of agitation because their diffusion in water is negligible. Additionally, agitation provides more collisions of the analytes with the fiber; therefore, it increases the sensitivity of analytes with high affinity for the fiber in a shorter extraction time. To illustrate, the exterior surface of the coated fiber is maintained at a constant analyte concentration that is determined by the analyte concentration in the aqueous phase adja-

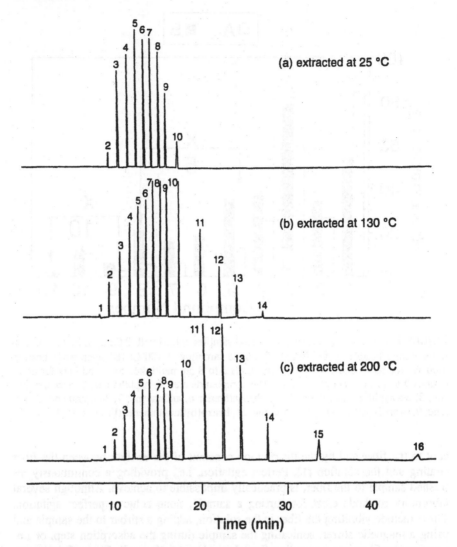

Figure 1.7 Total ion current chromatogram of 16 straight chain hydrocarbons sampled by headspace SPME from spiked sand at (a) 25°C, (b) 130°C, and (c) 200°C for 60 minutes. The components are: 1, C_{10}; 2, C_{11}; 3, C_{12}; 4, C_{13}; 5, C_{14}; 6, C_{15}; 7, C_{16}; 8, C_{17}; 9, C_{18}; 10, C_{20}; 11, C_{24}; 12, C_{28}; 13, C_{32}; 14, C_{36}; 15, C_{40}. (From Ref. 5)

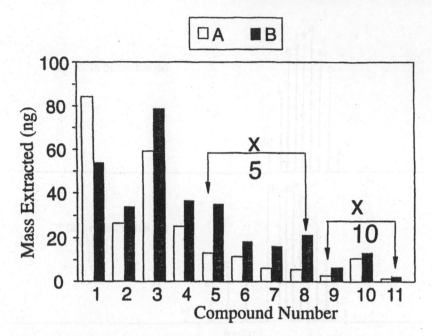

Figure 1.8 The mass extracted from sand samples spiked with 0.02 ppm PAHs and 5% water during headspace SPME: (A) 2-minute sampling at 100°C; (B) microwave heating (600 W) for 80 seconds. Bars for compounds 5 to 8 are multiplied by 5 and bars for compounds 9 to 11 are multiplied by 10. The compounds are: 1, naphthalene; 2, acenaphthylene; 3, acenaphthene; 4, fluorene; 5, phenanthrene; 6, anthracene; 7, fluoranthene; 8, pyrene; 9, benzo[a]anthracene; 10, chrysene; 11, benzo[b]fluoranthene. (From Ref. 5)

cent to the fiber and by the distribution constant of the analyte between the fiber coating and the solution [1]. Perfect agitation, i.e., providing a continuously refreshed sample to the fiber, is practically impossible to achieve. Although several laboratory methods exist for stirring a sample, none achieve perfect agitation. These include vibrating the fiber in the solution, adding a stirbar to the sample and using a magnetic stirrer, sonicating the sample during the adsorption step, or immersing the fiber into an on-line flow-through extraction cell. Each of these mechanical methods have both positive and negative practical aspects, which are discussed in Chapter 2; however, none provide the shorter equilibration times achieved with perfect agitation.

Figure 1.9 models the geometry of a direct extraction where the fiber is immersed into the liquid sample with no headspace present in the system. This model helps to explain the time required to extract the analytes from the solution. Here, the fiber coating outer radius, b, minus its inner radius, a, represents the fiber coating thickness.

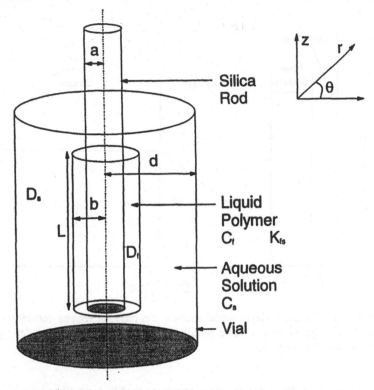

Figure 1.9 Graphic representation of the SPME/sample system configuration: a is fiber coating inner radius, b is fiber coating outer radius, L is fiber coating length, d is vial inner radius, C_f is analyte concentration in the fiber coating, D_f is analyte diffusion coefficient in the fiber coating, C_s is analyte concentration in the sample, D_s is analyte diffusion coefficient in the sample, K_{fs} is analyte distribution coefficient between fiber coating and sample, $K_{fs}=C_f/C_s$. (From Ref. 1)

For dynamic sampling, where the sample is stirred or agitated during sampling, a thin layer of water actually forms around the surface of the fiber coating, known as the Prandtl boundary layer (Figure 1.10) [5]. This layer is difficult to eliminate, even with rigorous stirring or agitating the fiber in the sample, although it is thinner than in the static sampling case which is discussed next. At farther distances from the fiber surface, the fluid movement gradually increases until it corresponds to the bulk flow in the sample. The agitation conditions and the viscosity of the fluid determine the thickness of the Prandtl boundary layer. Furthermore, if this layer is sufficiently thick, the diffusion of analytes through this layer determines the equilibration time [6]. Practically, the time required to reach equilibrium by stirring the sample or agitating the fiber is ten times less than under static sampling conditions.

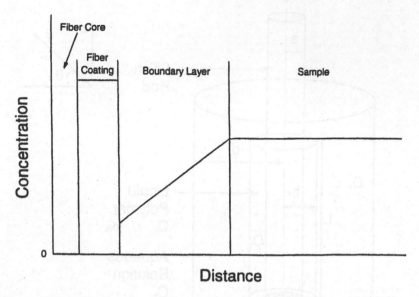

Figure 1.10 Boundary layer model showing the different regions considered and the assumed concentration versus radius profile for when the boundary layer determines the extraction rate. (From Ref. 5)

Figure 1.11 illustrates the time exposure profile for a fiber immersed in an agitated sample when the extraction rate is determined by the presence of a boundary layer [5]. The y-axis is the percentage of total analyte mass absorbed by the fiber at equilibrium. Time is represented in the x-axis by a dimension-free unit of the analyte diffusion coefficient in the sample fluid, D_s; divided by the boundary layer thickness, δ; the analyte's distribution constant between the fiber and sample, K_{fs}; the fiber coating thickness, (b-a); and the time to reach equilibrium. This time exposure profile clearly illustrates an initial, dramatic increase in the mass absorbed by the fiber that eventually evens out to reach equilibrium. Several interesting points can be learned about the extraction process from this graph. First, the time required to reach equilibrium is infinitely long. Nonetheless, a change in mass extracted cannot be determined if it is smaller than the experimental error, which is typically 5%. Consequently, one can practically assume that the equilibration time is achieved when 95% of the analyte's equilibrium amount is extracted from the sample. Second, the equilibration time is dependent on K_{fs}, the analyte's diffusion coefficient in the coating. Analytes with a high affinity for the fiber will have long equilibration times because more analytes must travel into the fiber. This is shown in Figure 1.6 for the BTEX components. Benzene, with the lowest K_{fs} of 125, takes approximately 250 seconds to reach equilibrium; whereas, p-xylene, with a K_{fs} of 831, takes 750 seconds. Moreover, the extraction time will be longer when using more cross-linked or thicker fiber coatings because

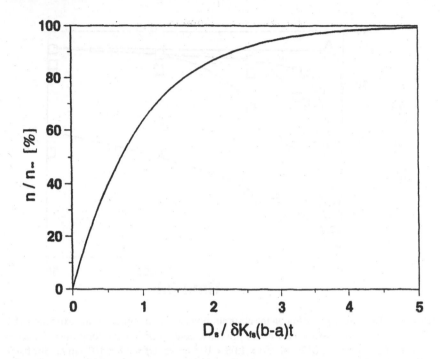

Figure 1.11 SPME time exposure profile showing the mass absorbed versus time from an agitated solution of infinite volume when the boundary layer controls the extraction rate. (From Ref. 5)

analytes will take longer to completely penetrate the coating (Figure 1.5). As previously mentioned, a thicker coating may be chosen to improve sensitivity because of its larger capacity; however, the compromise is a significant increase in extraction time. Finally, the equilibration time for an agitated solution with a boundary layer can be estimated from equation (1.7).

$$t_e = t_{95\%} = \frac{3\delta K_{fs}(b-a)}{D_s} \qquad (1.7)$$

Where: δ is the boundary layer thickness, K_{fs} is the analyte's distribution constant between the fiber and sample, $(b - a)$ is the fiber coating thickness, and D_s is the analyte diffusion coefficient in the sample fluid. This equation can be used to estimate the equilibration time when the extraction rate is controlled by the analyte's diffusion through the boundary layer. Of course, the more analytes that travel through the boundary layer, the longer the extraction process is.

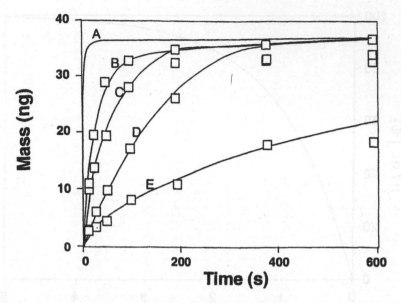

Figure 1.12 Effect of stirring on the absorption versus time profile for the extraction of 1 ppm benzene in water extracted with 56 μm thick coating on a 1 cm long fiber, for K_{fs} = 125, a = 0.007 cm, b = 0.0126 cm, Ds = 1.08 x 10^{-5} cm²/s, Df = 2.8 x 10^{-6} cm²/s. Stir bar 7 mm long, fiber positioned at center of vial, 7.4 mL vial. Curves are as follows: (A) theoretical prediction for perfect agitation; (B) 2,500 rpm; (C) 1,800 rpm; (D) 400 rpm; (E) no stirring. (From Ref. 5)

Figure 1.12 compares theoretical and experimental equilibration time profiles obtained by magnetic stirring at various rotational speeds for a benzene in water solution. Of course, the faster the stirring, the greater the mass is transferred to the fiber and, consequently, a shorter equilibration time [5]. Similarly, Figure 1.13 illustrates equilibration time profiles for different methods of agitating the same sample. Here, increased mass transfer is obtained by moving the fiber with respect to the aqueous solution (curve C) and by vibrating the vial in a sonicator bath (curve B), but neither are as efficient as a high rate of stirring (curve A).

Interestingly, the sample concentration has no affect on the concentration time profile and the equilibration time. Figure 1.14 demonstrates that when the extraction is optimized for a given concentration, the equilibration profile and time is the same for all other concentrations. This means that as long as the extraction method and distribution constants between the SPME/sample system remain constant, the system will behave linearly. Also note that a tenfold increase in response is achieved for every tenfold increase in Cs, the aqueous concentration, indicating a relatively infinite sample volume using equation (1.4).

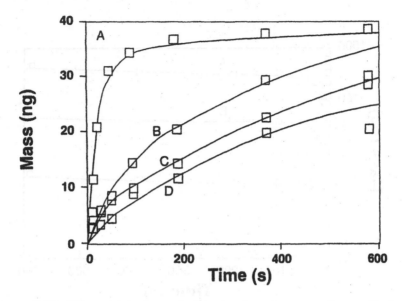

Figure 1.13 Effect of agitation method on the absorption versus time profile for the extraction of 1 ppm benzene in water. Parameters are the same as given for Figure 1.12. Curves are as follows: (A) 2,500 rpm magnetic stirring; (B) sonication with 1/8 inch horn disrupter tip placed in 50 mL vial at a low power less than 100 W; (C) manually repeated fiber insertion/retraction at a rate of one insertion and retraction per second; (D) no stirring. (From Ref. 5)

Direct Sampling from a Static Aqueous Solution

Static sampling, such as stagnant waters from ponds or ditches or sampling from an unstirred solution in the laboratory, has a much longer equilibration time due to low diffusion coefficients in water. Here, the mass transport of the analytes is limited by migration through the matrix surrounding the fiber surface. Naturally, the analytes closest to the fiber penetrate into it faster than the more distant analytes which can migrate through the water solution towards it. As a result, a static, aqueous layer, depleted of analytes, forms around the fiber (Figure 1.15). Initially, the rate of mass transport is high, but as the water layer around the fiber is depleted of analytes, diffusion through the water limits the rate of mass transport, and consequently impedes the extraction. Therefore, the final equilibration time is determined by diffusion through this progressively thicker, depleted layer [6]. Furthermore, the depleted water layer can greatly affect the extraction speed of analytes with high partition coefficients because more analytes must pass through the static water layer to reach the fiber coating [2]. Consequently, the more analytes that must be transported to the fiber, the longer the equilibrium process.

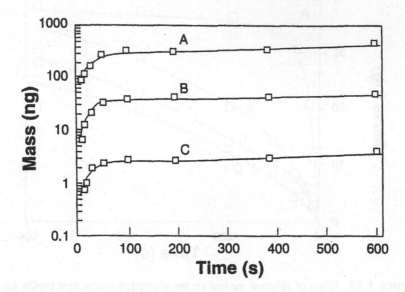

Figure 1.14 Effect of analyte concentration on the absorption vs. time profile of 2,500 rpm stirred benzene in water. Parameters are the same as given for Figure 1.12. (A) $C_s = 10$ ppm, (B) $C_s = 1$ ppm, (C) $C_s = 0.1$ ppm. (From Ref. 5)

Similar to the agitated case, the analyte concentration does not affect the equilibration time because the concentration gradients (and the flux) decrease proportionally with the decrease in analyte concentration. On the other hand, an increase in the distribution constant, i.e., the analyte affinity for the fiber, has a significant affect on increasing the extraction time if the sample is not perfectly agitated. The time for the larger distribution constant is increased because a larger amount of analyte must diffuse into the coating before equilibrium is reached. For example, Figure 1.16(a) compares the equilibrium times for four s-triazines and parathion when extracting under static sampling conditions. After 120 minutes, only simetryn with a low fiber coating/sample distribution constant reaches equilibrium. In this static case, the slow diffusion through the aqueous layer reduces the concentration gradient at the fiber coating surface, thereby limiting the analyte transport into the fiber (limiting the analyte flux). If similar equilibration times are to be achieved for all distribution constants, then a proportional increase in the analyte flux is required for the larger distribution constants. This higher flux of analyte into the fiber coating is accomplished by agitating the sample, which increases the concentration gradient at the coating/solution interface. Graph (b) illustrates the results from stirring the sample. Here, all five compounds reach equilibrium within 50 minutes using this classic method for achieving faster mass transport into the fiber. Finally, the results of vibrating the fiber using an autosam-

Figure 1.15 Depleted area around the fiber during static sampling of semivolatiles. (Courtesy Z. Penton, Varian Associates, Inc.)

pler show that equilibrium was achieved within 35 minutes for all compounds (c). These results are obtained using 2 mL vials; however, Chapter 2 will compare extraction times using larger volumes.

Speed of Extraction for Headspace Sampling

Headspace SPME sampling time is dependent on the kinetics of mass transport in which the analytes move from the aqueous phase to the headspace and finally to the fiber coating. To illustrate, an aqueous sample contaminated with organic compounds is transferred to a closed container with headspace. Then, equilibrium is established between the aqueous solution and the headspace. The SPME fiber, coated with a thin layer of a selected liquid organic polymer, is inserted into the headspace of the container; therefore, the fiber does not have any direct contact with the aqueous phase. Figure 1.17(a) illustrates the SPME headspace model. As the fiber's liquid coating absorbs organic analytes, the analytes undergo a series of transport processes from the water to gas phase, and eventually

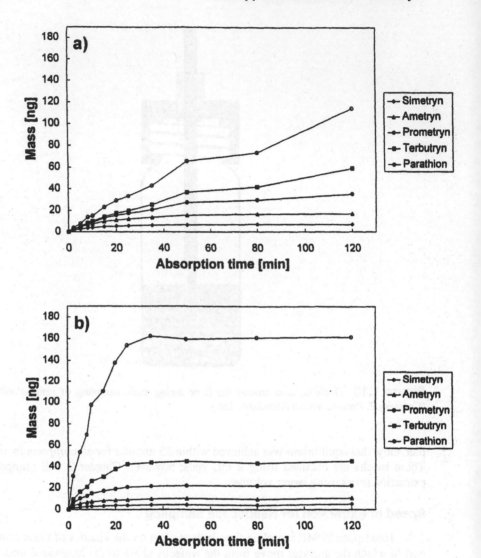

Figure 1.16 Absorption time profiles for four s-triazines and parathion using (a) static absorption conditions, (b) magnetic stirring, and (c) fiber vibration method (for 2-mL vials). PDMS, 100 μm fiber used. Equilibrium is reached substantially faster for agitation techniques. (From Ref. 6)

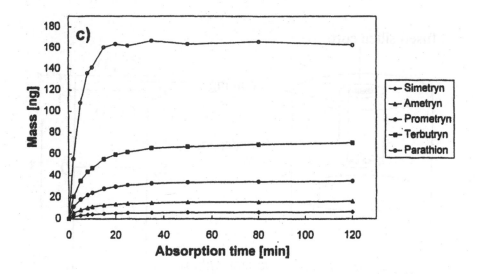

to the coating, until the system reaches equilibrium. The extraction equilibrium is reached when the analyte concentration is homogenous within each of the three phases, and the concentration differences between two adjoining phases have satisfied the values of their partition coefficient. That is, when equation (1.1) is satisfied. Of course, the release of volatile analytes into the headspace is relatively easy because the compounds tend to vaporize once they are dissociated from their matrix. For semivolatile compounds, the low volatility and relatively large molecular size may slow the mass transfer from the matrix to the headspace, consequently resulting in a long extraction time. Stirring the solution continually generates a fresh, aqueous surface which helps the less volatile components to escape to the headspace. In this discussion, both stirred and unstirred aqueous solutions will be examined.

Headspace Sampling from a Static Aqueous Solution

First, let's examine the effect of varying the fiber coating/headspace partition coefficients (K_{fh}) when the headspace/sample partition coefficient, K_{hs}, is held constant in an unstirred solution. Similar to direct sampling, more analytes are absorbed by the fiber coating and more time is required to reach the final equilibrium for compounds with a greater affinity for the fiber. The longer equilibrium time is caused by the very slow diffusion of analyte molecules in the aqueous phase. Even though the diffusion coefficient in the fiber coating is actually less

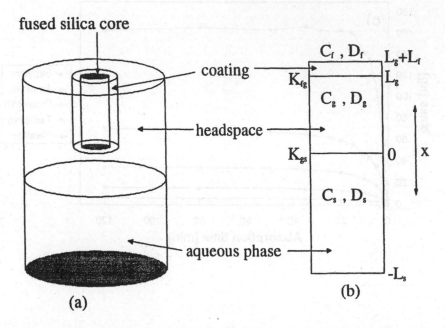

Figure 1.17 (a) Geometry for SPME headspace sampling. (b) One dimensional model of the three-phase system; K_{fh} and K_{hs} are the coating/headspace and headspace/water partition coefficients; D_f, D_h, and D_s are the diffusion coefficients of the analyte in the coating, headspace, and water; C_f, C_h, and C_s are the concentrations in the coating, headspace, and water; L_f, L_h, and L_s are the thicknesses of coating, headspace, and aqueous phases. (From Ref. 2)

than in the aqueous phase, the coating thickness is considerably less than the other two phases; consequently, the diffusion within the coating consumes little time.

On the other hand, if the analytes' affinity for the fiber is equal, but their volatility differs, of course the more volatile analytes will have greater chance to be absorbed by the fiber. Therefore, the equilibrium times are shorter and more analyte is absorbed by the fiber due to its greater concentration in the headspace. When the analyte has a small K_{hs} value, the analyte concentration in the headspace is low, the headspace extraction affects the aqueous phase concentration, and equilibrium takes a long time to establish itself. Consequently, the best headspace sampling results from volatile analytes (high K_{hs} values) that will transport more molecules in the headspace, especially without the aid of stirring the sample. Of course, the sample temperature can be a significant experimental parameter to optimize during extraction. When the temperature is high enough to move the analytes from the sample matrix to the headspace, the equilibrium time can be reduced; moreover, it is independent of the agitation conditions [1].

Headspace Sampling from an Agitated Aqueous Solution

Aside from heating the sample, constantly stirring the water sample to generate a continuously fresh surface will also accelerate the mass transfer of less volatile analytes from the water to headspace. Once in the gaseous phase, analytes transfer from the headspace to fiber coating rapidly because of their large diffusion coefficients in the gas phase. Additionally, volatile compounds transfer more efficiently from the water to gas to fiber coating than from the water to fiber coating. Similar to the static headspace sampling conditions, a group of compounds with similar volatility will demonstrate longer equilibration times for the analytes that have a greater affinity for the fiber. However, the equilibration in this case is longer because more analytes need to be transported through the gas phase, which is now the limiting phase [5].

Also similar to the non-agitated case, the equilibration time is longer for the less volatile analytes in a group of compounds that have similar affinities for the fiber [5]. Because the aqueous phase is well-agitated and the fiber coating is very thin, the limiting step is now diffusion through the headspace, i.e., from the headspace/water interface to the coating/headspace interface.

In summary, stirring does not affect analyte diffusion from the headspace to the fiber coating. However, if exposing the SPME fiber in the headspace significantly reduces the headspace concentration, then the mass transport between the

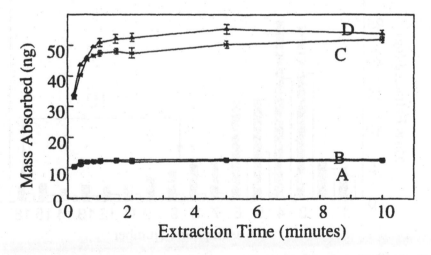

Figure 1.18 The extraction time profiles of 1 ppm benzene and *o*-xylene in aqueous solutions during headspace SPME: (A) benzene with a static aqueous phase; (B) benzene with a well-agitated aqueous phase; (C) *o*-xylene with a static aqueous phase; (D) *o*-xylene with a well-agitated aqueous phase. Benzene: $K_{fh} = 493$, $K_{hs} = 0.26$; *o*-xylene: $K_{fh} = 4417$, $K_{hs} = 0.15$. (From Ref. 2)

aqueous sample and the headspace slows the extraction process. Stirring accelerates the equilibrium between these two phases during sampling by continually replenishing the headspace concentration. To illustrate, Figure 1.18 [17] shows the time profile of mass absorbed by a fiber versus sampling time with and without mixing. Note that benzene, which has a relatively small K_{fh} value (affinity for the fiber coating) and large K_{hs} value (volatility), has almost identical extraction profiles for both unstirred (curve A) and stirred (curve B) conditions. Agitation has little affect on highly volatile compounds with relatively small fiber/headspace partition coefficients because of these compounds' large headspace capacity. On the other hand, agitation provides a significant improvement in equilibration for *o*-xylene, which has a large affinity for the fiber, but low volatility.

Finally, Figure 1.19 demonstrates that sonication is more effective than stirring for headspace sampling of polyaromatic hydrocarbons (PAHs) in water. Sonication yields a greater response for each component, indicating that the more aggressive agitation method forces the components into the headspace where they can interact with the fiber.

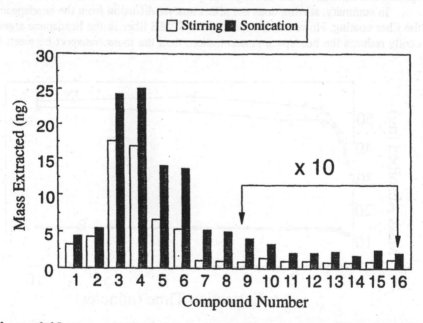

Figure 1.19 The mass of PAHs extracted by the fiber coating from aqueous samples of 0.1 ppm PAHs using headspace SPME using either sonication or magnetic stirring; bars from compounds 9 to 16 were enlarged 10 times. The compounds are: 1, naphthalene; 2, acenaphthylene; 3, acenaphthene; 4, fluorene; 5, phenanthrene; 6, anthracene; 7, fluoranthene; 8, pyrene; 9, benzo[a]anthracene; 10, chrysene; 11, benzo[b]fluoranthene; 12, benzo[k]fluoranthene; 13, benzo[a]pyrene; 14, indo[1,2,3-cd]pyrene; 15, dibenzo[a,h] anthracene; 16, benzo[ghi]perylene. (From Ref. 5)

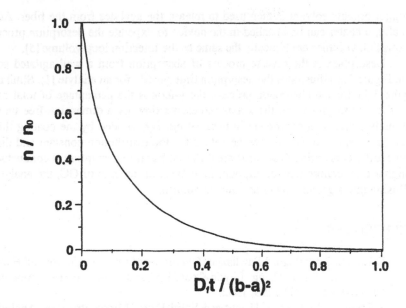

Figure 1.20   Mass in the fiber versus time, showing the mass desorbed from the fiber by a high velocity stripping phase (carrier gas or mobile phase) after a fiber is exposed in an instrument injection port. (From Ref. 1)

DESORPTION

Once the analyte extraction is complete, the coated fiber containing the analytes is ready to be desorbed into a GC or HPLC injector for analysis. The extraction step establishes the more critical experimental parameters (speed, sensitivity, accuracy, and precision). Desorption, which is closely related to the efficiency of the chromatographic separation and the precision of quantitation, has a substantial influence on the quality of data obtained [4]. For GC analysis, the desorption process involves inserting the fiber into a hot GC injector. As the temperature increases, the coating/gas partition coefficients decrease and the fiber coating's ability to retain analytes quickly diminishes. Additionally, the constant flow of carrier gas within the GC injector helps discharge the analytes from the fiber coating and transfers them to a cool column for refocusing. Typically, a desorption time of two minutes is adequate to release all analytes from the fiber coating. On the other hand, desorption using solvents and subsequent HPLC analysis is preferred for nonvolatile and thermally labile analytes. In this case, the SPME fiber is inserted into a desorption chamber which replaces the injection loop on the injector valve. When the valve is in the "load" position, the fiber is introduced into the desorption chamber under ambient pressure. The mobile phase or a stronger organic solvent is then used to release the analytes from the fiber. Additionally, a heater can be installed in the device to expedite the desorption process. Desorption volumes are typically the same as the injection loop volume [5].

Desorption is the reverse process of absorption from a well-agitated solution. Figure 1.20 illustrates the desorption time profile for an analyte [1]. Similar to Figure 1.11 for the absorption process, the y-axis is the percentage of total mass absorbed by the fiber and the x-axis represents time by a dimension-free unit of the analyte diffusion coefficient in the coating, $D_f t$, divided by the coating thickness, $(b-a)$, squared. Of course, the values for the distribution constant and diffusion coefficients during desorption are different because desorption is conducted at a higher temperature than absorption; moreover, in the case of GC, the analyte is diffusing into a gas instead of an aqueous solution.

REFERENCES

1.   D Louch, S Motlagh, J Pawliszyn. Dynamics of Organic Compound Extraction from Water Using Liquid-Coated Fused Silica Fibers. Analytical Chemistry 64:1187-1199, 1992.
2.   Z Zhang, J Pawliszyn. Headspace Solid-Phase Microextraction. Analytical Chemistry 65:1843-1852, 1993.
3.   CL Arthur, J Pawliszyn. Solid Phase Microextraction with Thermal Desorption Using Fused Silica Optical Fibers. Analytical Chemistry 62:2145-2148, 1990.
4.   Z Zhang, MJ Yang, J Pawliszyn. Solid-Phase Microextraction. Analytical Chemistry 66:844-853, 1994.
5.   J Pawliszyn. Solid Phase Microextraction Theory and Practice. New York: Wiley-VCH, 1997, pp. 37-94.
6.   R Eisert, J Pawliszyn. Design of automated solid-phase microextraction for trace analysis of organic compounds in aqueous samples. J. Chromatography A 776:293-303, 1997.

2

Method Development with Solid Phase Microextraction

Zelda Penton
Varian Chromatography Systems, Walnut Creek, California

INTRODUCTION

Of all the steps in a GC analysis, sample preparation is probably the area where the most confusion exists. For many samples, Solid Phase Microextraction (SPME) can replace a tedious, multi-step sample preparation technique; therefore, the number of publications by enthusiastic users has significantly increased. In my function as an application chemist, I encounter people who are aware of the increasing use of SPME and wonder if it might be appropriate for their analytical problems.

This chapter will begin with a comparison of SPME with other sample preparation techniques to help the reader determine if SPME is appropriate for a particular analysis. This will be followed by a step-by-step procedure for getting started with SPME. Special considerations for achieving quantitative results will be covered and the chapter will conclude with a trouble-shooting section. An assumption is made here that the SPME cleanup will be followed by GC or GC/MS analysis; SPME/HPLC is discussed in Chapter 3.

EVALUATING SPME VERSUS OTHER SAMPLE PREPARATION TECHNIQUES

Some guidelines will be given here to help the user decide if a sample preparation procedure is necessary and if SPME is appropriate for a particular analysis. Fortunately, for those samples that fall into a gray area, SPME can be evaluated in a short time at a relatively reasonable cost.

Begin with the following questions:

1. Is the sample "dirty" (for example, does it contain non-volatile material such as heavy oils, tissue, protein, or undissolved solids?)?
2. Is it in a water-based matrix?
3. Are the compounds to be determined at very low levels necessitating concentration steps?

If the answer to any or all of these questions is "yes," then some sort of sample preparation is necessary.

The ideal candidate for SPME is a relatively clean, aqueous sample containing volatile or semivolatile organic compounds. If all of the analytes are volatile (b.p. <150°C), the matrix can be relatively dirty; i.e., non-volatile compounds such as proteins and particulates can be tolerated. Analyte concentrations can be in the low ppb to the low percentage range. SPME is also suitable for monitoring outgassing volatiles in solid samples. Virtually all samples that can be analyzed by conventional static headspace can be determined with SPME.

Semivolatiles in dirty matrices are more difficult to determine with SPME. They require heating to drive the analytes into the headspace or immersion of the SPME fiber, followed by rinsing to remove non-volatiles prior to injection. Reports in the literature [1,2] describe polar compounds determined by SPME with derivatization on the fiber.

For many samples, cleanup with SPME can replace Purge and Trap, solid phase extraction, and liquid-liquid extraction. Tables 2.1 to 2.3 compare some of the features of SPME with those of other sample concentration methods.

Table 2.1 Comparing Features of Solid Phase Extraction with SPME.

Feature	Solid Phase Extraction	SPME
Sample matrices	Liquids—organic or aqueous and some semi-solids such as tissue.	Aqueous liquids, solids, gas (rarely).
Sample pretreatment	Blood, tissue and some environmental samples usually require pretreatment to remove solids or proteins; many liquid samples do not.	Sample pretreatment usually not required for volatiles. Analytes not amenable to headspace sampling may require some matrix cleanup.
Analytes	Semivolatiles and slightly volatile compounds; recovery a problem with volatiles.	Volatile and semivolatile compounds.
Use of organic solvents	Environmental samples: typically 25 mL to extract one liter, versus 300-600 mL with liquid-liquid extraction. Biological samples: 4-8 mL (same as liquid-liquid extraction)	None.
Recovery of analytes	Generally, recoveries close to 100% are expected.	Equilibrium method—quantitation usually by comparing to spiked blank matrix or by standard additions.
Automation	Limited on-line for GC.	On-line available.
Government regulatory agency approval	Several approved methods.	Relatively new technique (no approved methods at present).
Advantages	Better sensitivity because full recovery is possible from samples as large as one liter.	Much simpler. Usually the user places samples in vials and starts absorbing onto fiber.

Table 2.2 Comparing Features of Static Headspace with SPME.

Feature	Static Headspace	SPME
Sample matrices	Aqueous liquids and solids.	Aqueous liquids, solids, gas (rarely).
Sample pretreatment	None for many samples.	None for many samples.
	Matrix modifiers sometimes used such as salting out or adjusting pH.	Matrix modifiers sometimes used such as salting out or adjusting pH.
Analytes	Volatile compounds (up to approximately 300°C).	Volatiles and semivolatiles.
Use of organic solvents	None.	None.
Recovery of analytes	Equilibrium method—quantitation usually by comparing to spiked blank matrix or by standard additions.	Equilibrium method—quantitation usually by comparing to spiked blank matrix or by standard additions.
Automation	Normally used with automation.	Automation available.
Government regulatory agency approval	Several approved methods.	Relatively new technique (no approved methods at present).
Advantages	Older, more established method.	Comparable recovery to SHS with most volatiles but better recovery with higher boiling compounds such as xylenes, simple PAH's with most fibers.
		No active transfer lines.
		Lower capital cost for dual function hardware (liquid injection and SPME).
		Requires no bench space.

Table 2.3 Comparing Features of Purge and Trap with SPME.

Feature	Purge and Trap	SPME
Sample matrices	Aqueous liquids and solids	Aqueous liquids, solids, gas (rarely)
Sample pretreatment	None for many samples.	None for many samples.
	Matrix modifiers sometimes used such as salting out or adjusting pH.	Matrix modifiers sometimes used such as salting out or adjusting pH.
Analytes	Volatile compounds (up to approximately 300°C)	Volatiles and semivolatiles
Use of organic solvents	None	None
Recovery of analytes	Recoveries up to 100% expected	Equilibrium method—quantitation usually by comparing to spiked blank matrix or by standard additions.
Automation	Normally used with automation	On-line available
Government regulatory agency approval	Several approved methods	Relatively new technique (no approved methods at present).
Advantages	Greater sensitivity (~100 fold) expected for purgeable compounds.	Lower capital cost for dual function hardware (liquid injection and SPME) and lower costs for maintaining and troubleshooting hardware.
		Requires no bench space.
		No problems with foaming samples.

SPME versus SFE

Another sample preparation technique, Supercritical Fluid Extraction (SFE) is normally used to extract volatiles and semivolatiles from solids and semi-solid samples, such as foods, sludges and soils. In some situations, SPME may be useful for extracting compounds in the headspace above such samples but semivolatiles in very viscous organic matrices cannot be extracted by SPME.

Comparison of Results with Volatiles

An aqueous sample containing volatiles was analyzed after extraction with SPME, static headspace, and Purge and Trap. Detection limits and precision for all of these techniques is compared with direct injection of the liquid sample in Table 2.4.

In some cases, where SPME is not the technique of choice for sample cleanup, it may be useful as a screening procedure. For example, the use of Purge and Trap is presently mandated by the U.S. EPA for determining volatiles in drinking and wastewater. SPME has been used for initial screening of samples to determine if dilution is necessary prior to the Purge and Trap procedure, thus eliminating the overloading of the trap and reducing maintenance costs [3].

Table 2.4 Detection Limits and Precision of Organic Volatiles[a] in Water.

Technique	Detection Limit with FID (ppb)	Precision (% rsd)
SPME	0.1-10	1-3
Static headspace (heated)	0.1-2	1-3
Purge and Trap	0.003-0.05	1-8
Direct Injection	17-240	2-13[b]

[a]methylene chloride, chloroform, dioxane, TCE, benzene, toluene, xylene, and 1,2,4-trimethylbenzene.
[b]Better results are usually obtained when an organic solvent, rather than water, is injected.

PROCEDURE FOR SAMPLE PREPARATION WITH SPME

Preparation of the GC

The assumption here is that chromatographic conditions have been optimized for the analytes, i.e., an appropriate column, temperature program, and detector have been selected. Sample introduction with SPME does not preclude any of the normal GC techniques. Cryogenics in the column oven are advantageous when the sample contains analytes with a wide range of volatility (for example, vinyl chloride and PAH's); this is generally true, and is not a characteristic of SPME sampling. Often, cooling the column to sub-ambient temperatures can be avoided by selecting a GC column with a film thickness of one micron or greater.

Any detector including a mass spectrometer can be used successfully. Moreover, various techniques used for confirmation, such as splitting the sample to two columns or introducing an effluent splitter at the end of the column, will be unaffected by SPME introduction. Only the injector conditions should be optimized for SPME.

Injector Septum

The SPME fiber assembly includes a septum-piercing needle, which is a blunt, hollow 23-24-gauge[1] tube. In comparison, liquid injection into a GC is usually accomplished with a tapered 26-gauge[2] needle. Therefore, sample introduction with an SPME fiber is more likely to result in septum failure. A septumless injector seal such as those manufactured by Merlin Instruments, Half Moon Bay, CA, 94019, or Jade Systems, Austin, TX, 78753, is recommended (Figure 2.1).

---

[1] Originally, Supelco used 24-gauge tubing in manufacturing SPME fibers, but 23-gauge tubing was available for septumless seals such as the Merlin microseal. By the time this work is published, it is expected that 23-gauge tubing will be standard.

[2] The higher the gauge number, the narrower the outer diameter.

Figure 2.1 The Merlin Microseal can be installed in a GC injector in place of a septum. The device contains a "duckbill" that allows a needle to enter the injector without leaking. (Courtesy of Varian Associates, Walnut Creek, CA 94598.)

These products are not generic; they are designed for specific GC injectors. Another alternative is to use a pre-drilled septum; however, these tend to be very difficult to seal. Nevertheless, use of a conventional GC septum is possible with SPME. Figure 2.2 shows no change of retention time after 46 runs, indicating that the septum is intact. Frequently, a septum lasts 100 runs or more without problems. To minimize septum failure, the following procedure is recommended:

1. Install a new septum.
2. Puncture the septum with an SPME sheath three or four times. With automation, lower the AutoSampler carriage manually so that the septum-piercing needle penetrates the septum several times.
3. Remove and inspect the new septum. Pull off and discard any loose particles of septum material.
4. Reinstall the septum.

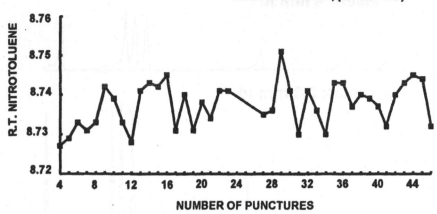

Figure 2.2 Consistent retention times are achieved when an SPME fiber punctures a conventional GC injector septum 46 times. Occasionally, problems can occur (fiber breakage/ghost peaks) when corings from the septum enter the injector insert. (Courtesy of Varian Associates, Walnut Creek, CA 94598.)

    With any of the foregoing methods, the user should monitor the head pressure on the column as the fiber sheath enters and leaves the injector, to verify the integrity of the seal. A subtle leak will be indicated by unusual shifts in retention time or the presence of air in a mass spectrometer.

    The injector insert is a significant factor in assuring good results when an SPME fiber is desorbed. While it is obvious that the insert should not be packed or contain any physical obstructions when the fiber is pushed out of the septum piercing needle, it is also necessary to consider the inner diameter of the insert. This should be between 0.75 to 0.80 mm. An insert of smaller diameter will not allow the fiber sheath to penetrate the injector. Larger inserts (2-4 mm id) will result in the broadening of early-eluting peaks (Figure 2.3).

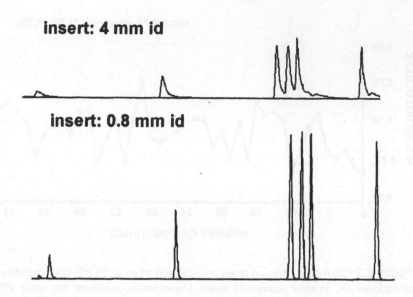

insert: 4 mm id

insert: 0.8 mm id

Figure 2.3 SPME chromatograms of BTEX's after injection into two inserts of different internal diameters. (Courtesy of Varian Associates, Walnut Creek, CA 94598.)

Injector inserts for SPME are available commercially and may be used for split or splitless injection. With splitless injection, the vent is timed to open at the end of the desorption period (usually 2-10 minutes).

Injector Temperature

SPME fibers are generally desorbed under hot, isothermal conditions; although, temperature-programmable injectors are popular for minimizing decomposition of labile compounds and eliminating discrimination based on volatility. Rapid desorption from the fiber is necessary for sharp peaks [4] without sample carryover. Injector temperature is normally 10-20°C below the temperature limit of the fiber and/or the GC column (usually 200° to 280°C).

Automation

With SPME, the absorption and desorption times must be consistent to achieve good precision. This requires the analyst to remain in the area of the analysis; consequently, manual SPME can be quite time-consuming. Varian Chromatography Systems, Walnut Creek, CA 94598, manufactures an SPME autosampler that is controlled by a PC and can be installed on any gas chromatograph. The autosampler was originally designed for liquid injection and can be used in either SPME or liquid injection mode. For laboratories where more than a few samples per week are to be analyzed with SPME, the autosampler will greatly improve productivity and should improve precision. The autosampler allows the sample to be mixed and heated, if desired. An especially useful feature is automatic method download, which allows the user to examine the effect of changing various parameters automatically during the method development phase. Thus, the effect of varying sampling times, mixing, etc., can be studied while the analyst is free for other tasks.

Selection of the Fiber, Sampling Vial, and Vial Septa

Supelco, Inc., Bellefonte, PA, 16823, manufactures the SPME fibers and holders, and they are continually improving fiber technology. Presently, five phases are commercially available for GC injection. The first phase introduced commercially was polydimethylsiloxane (PDMS). This phase is particularly suitable for non-polar compounds and is available in three film thicknesses, with the thinner films recommended for samples with less-volatile analytes. More recently, phases for polar and very volatile compounds were introduced. Table 2.5 summarizes the fibers that are commercially available.

Table 2.5 Commercially Available SPME Fibers for GC and GC/MS.

Phase	Applications
Polydimethylsiloxane Three film thicknesses are available: • 7 µm • 30 µm • 100 µm	Non-polar phase (for many semipolar compounds: aromatics, esters, many pesticides). 100 um used for relatively volatile compounds; the thinner phases are for non-polar and semipolar compounds of low volatility.
85-µm Polyacrylate	Polar compounds such as phenols, esters.
65-µm Carbowax/divinylbenzene	More polar than polyacrylate, for alcohols.
75-µm PDMS/divinylbenzene	Moderately polar, for amines.
65-µm Carboxen/PDMS	Highly volatile compounds including vinyl chloride, sulfur gases.

For samples with a mixture of different classes of analytes, the user should compare two or three fibers and consider which phase gives the best overall performance. Sensitivity is not the only issue, other considerations are a relative lack of affinity for interfering compounds and the ability of the fiber to be desorbed easily so that no sample carryover occurs. Fiber selection is covered thoroughly in Chapter 3.

Minimizing Interferences in Blank Runs

In all GC analyses, the analyst should make every effort to minimize the appearance of extraneous peaks to assure precise and definitive results with maximum sensitivity. With conventional liquid injection, these peaks are usually derived from contamination in the injector, septum bleed, column bleed, or from impurities in the solvent. Many of these problems have been reduced in the past few years with improvements in column manufacturing technology, septa designed for high temperature applications, capillary injectors with septum purge lines, and highly purified solvents.

SPME is an extremely useful technique because the fibers are capable of extracting and concentrating trace amounts of organic contaminants and the new user should be aware that extraneous peaks can be a significant problem. The method has existed for a relatively short time; therefore, manufacturers have not yet addressed some of the sources of contamination. It is well known among SPME users that a new fiber must be conditioned before use; the problem is that large interfering peaks can still occur with blank runs of a well-conditioned fiber. This is illustrated in Figure 2.4.

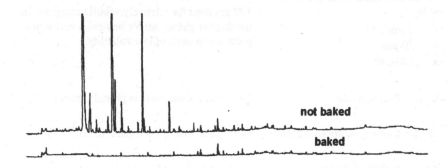

Figure 2.4 SPME sampling of empty 2-mL vials with a well-conditioned 100-μm polydimethylsiloxane fiber. The top chromatogram is the result of sampling a vial with a septum that was used as shipped by the manufacturer. The bottom chromatogram was obtained after the septum was baked for several hours at 150° C. (Courtesy of Varian Associates, Walnut Creek, CA 94598.)

The GC system used in the study was quite clean, and no ghost peaks occurred when the column was temperature-programmed without injection. When the fiber sampled a blank vial, many peaks resulted; moreover, some of them were quite large. Most of these peaks were identified by GC/MS as siloxanes and were found to be derived from the vial septum. Vial septa from the same manufacturer and from the same lot gave very consistent patterns; furthermore, some brands of septa were much worse than others. Most of the contamination was removed by baking the vial septa overnight in a laboratory oven at 150°C. The remaining extraneous peaks were identified as nitrogen-containing residues from the glue used to hold the SPME fiber in the fiber assembly.

Hopefully, as SPME matures, sampling vials that are certified to be free of siloxanes and other interfering material will become available, and the fibers, themselves, will be free of contamination. The user should also be aware that contamination may be found in water that is used for making standards. Bottled drinking water is often cleaner for SPME purposes than reagent water certified for use in HPLC applications.

The extent of the contamination problem with SPME varies with the concentration levels of the analytes of interest. Fortunately, it can often be minimized by the use of selective GC detectors, by selective ion monitoring with mass spectrometry, and by the use of fibers that have little affinity for the particular interferent. An example of the latter would be determining alcohols with a Carbowax/divinylbenzene fiber because these fibers have little tendency to absorb siloxanes.

Optimization of SPME Sampling Conditions

Effect of Various Parameters on Sensitivity

Intuitively, the following techniques should be significant in maximizing SPME sensitivity. The effectiveness of these measures, listed below, will be discussed next.

1. Sampling the liquid rather than the headspace
2. Mixing the sample during absorption
3. Changing the partition coefficient between the fiber and the other phases so that analytes are driven into the fiber
4. Saturating the sample with salt
5. Heating the sample during SPME sampling
6. Maximizing the ratio of liquid to headspace volumes in the vials
7. Using larger sampling vials

Sampling the Liquid Rather than the Headspace

With liquid samples, the recovery expected from SPME is dependent on the partitioning of the analytes between the three phases present in the sampling vial—the liquid (normally aqueous), the headspace above the liquid, and the fiber coating[3].

The equations governing this equilibrium process are:

$K_{fh} = \dfrac{C_f}{C_h}$ Where: K_{fh} is the partition coefficient of an analyte between the fiber coating and headspace phases, and C_f and C_h are the concentrations of the analyte in these phases.   (2.1)

$K_{hs} = \dfrac{C_h}{C_s}$ K_{hs} is the partition coefficient of an analyte between the headspace and aqueous phases, and C_h and C_s are the concentrations of the analyte in these phases.   (2.2)

$K_{fs} = \dfrac{C_f}{C_s}$ K_{fs} is the partition coefficient of an analyte between the fiber coating and aqueous phases, and C_f and C_s are the concentrations of the analyte in these phases.   (2.3)

The distribution among the three phases after equilibrium is represented as:

$$C_o V_o = C_h V_h + C_s V_s + C_f V_f \qquad (2.4)$$

Where: C_o is the concentration of the analyte in the original liquid sample, and V_s, V_h, and V_f are the volumes of the liquid, headspace, and fiber phases. Equations 2.1 to 2.4 can be rearranged to equation 2.5 from which one can calculate the effect of changing certain variables, such as the relative volumes of phases.

$$C_f = \frac{C_o V_s K_{fs}}{V_f K_{fs} + K_{hs} V_h + V_s} \qquad (2.5)$$

For example, if the partition coefficients K_{hs} and K_{fs} have been previously determined, equation 2.5 can be used to predict the concentration of an analyte in the fiber. This will enable the analyst to determine recovery of the compound from the matrix. These calculations are useful when equilibrium exists between the three phases.

---

[3] With solid samples, the partitioning is also between three phases but equilibrium is often not achieved between the solid phase and the headspace.

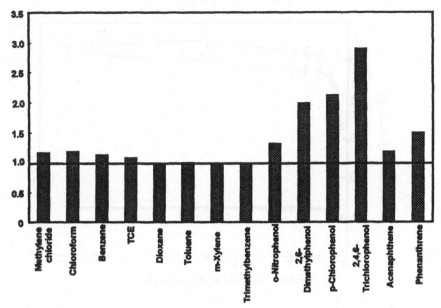

Figure 2.5 Response ratios for various compounds comparing SPME sampling of the liquid phase versus the headspace. Note, if the ratio is 1.0, the response is the same. Absorption time is 10 minutes, concentrations vary from 0.2-0.4 ppm in water, pH 2.0. (Courtesy of Varian Associates, Walnut Creek, CA 94598.)

After equilibrium is attained, the concentration of analytes in the fiber should be the same whether the fiber is immersed in the liquid or the headspace [5]. Figure 2.5 compares the relative responses of compounds in a test mixture after SPME sampling of the liquid and the headspace. Practically, liquid sampling is necessary for compounds of very low volatility, such as pesticides. Semivolatile compounds diffuse relatively slowly into the headspace above the liquid phase and into the fiber; therefore, many hours of absorption times are required to achieve a satisfactory response with headspace sampling.

In summary, SPME headspace sampling will yield detection limits comparable to SPME liquid sampling for volatiles but not for semivolatiles. Determining semivolatiles in "dirty" samples is discussed later in this chapter.

Mixing the Sample during Absorption

Mixing the sample is effective in increasing the response for compounds of low volatility, but is not necessary for volatiles. Figure 2.6 is a curve of mass absorbed by a fiber versus sampling time, with and without mixing. Note that with benzene, a volatile compound with a high diffusion rate, equilibrium and maximum response is achieved in less than one minute, regardless of whether the sample is mixed.

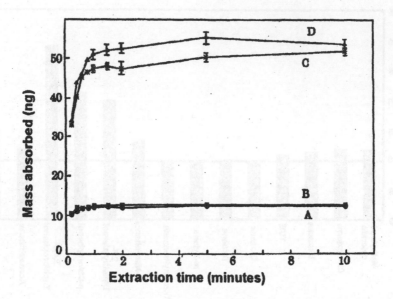

Figure 2.6 Time profile of the mass absorbed by the fiber coating with an aqueous solution of 1 ppm initial concentration of benzene and o-xylene: (A) benzene with a static aqueous phase; (B) benzene with a well-agitated aqueous phase; (C) o-xylene with a static aqueous phase; (D) o-xylene with a well-agitated aqueous phase. (From Ref 5).

With o-xylene, mixing enhances the response approximately 10% after two minutes of sampling. With compounds of low volatility, such as polyaromatic hydrocarbons (PAH's), mixing has a significant effect, whether the headspace or the liquid is sampled. With acenaphthene, equilibrium was attained in 30 minutes with stirring, but the equilibrium time was reduced to 10 minutes when the rate of stirring was increased by one-third. Arthur et al. [6] explain this phenomenon by suggesting the existence of a depleted area around the fiber, due to slow diffusion of relatively large molecules in the liquid phase.

Changing the Partition Coefficient between the Fiber and the Other Phases
This is done to drive the analytes into the fiber, and is normally achieved by changing to a fiber phase with better affinity for the analytes (Chapter 3).

Saturating the Sample with Salt
Partitioning of organic analytes out of an aqueous phase can be affected by changing the composition of the liquid phase. The addition of salt to aqueous samples is frequently used to drive polar compounds into the headspace [7], but it has a relatively insignificant effect on non-polar compounds (Figure 2.7).

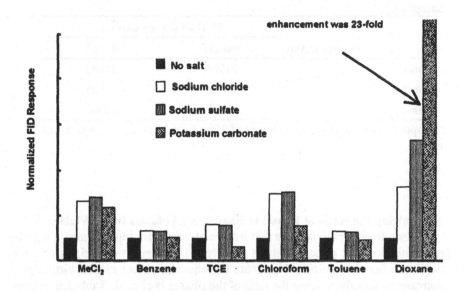

Figure 2.7 The effect of adding various salts to volatiles in water (1 ppm each) on the concentration in the headspace. "Salting out" has the most effect with dioxane, the most polar compound. Generally, for polar compounds, salts with multivalent ions will have a greater effect on altering the partition coefficients than monovalent salts. (Courtesy of Varian Associates, Walnut Creek, CA 94598.)

Heating the Sample during SPME Sampling

Although heating the sample is often useful to enhance sensitivity in static headspace, it is somewhat less effective when sampling volatiles with SPME. The difference arises from the fact that with SPME, three phases exist, and although heating tends to drive analytes out of the liquid phase and into the headspace, heating also alters the partitioning of the analyte between the headspace and the fiber to favor the headspace. Zhang and Pawliszyn [8] have shown that sensitivity can be enhanced significantly by heating the sample and simultaneously cooling the fiber with liquid CO_2.

In a laboratory with wide fluctuations in temperature, thermostatting would be expected to improve precision. In such cases, an internal standard might have the same effect. Obviously, when sampling at elevated temperatures, it is particularly important to use vials that will not leak and lose analyte with the increased pressure that will be generated. Crimp-top vials tend to be superior to screw-cap vials at higher temperatures.

Table 2.6 Effect of Sample Volume on Methanol Area Counts after SPME Sampling.[a]

| | Methanol Area Counts | |
| | 200 µL | 600 µL |
Volume in vial		
run 1	22597	21645
run 2	22622	22095
run 3	21877	22306

[a] Response was the same for volumes of 200 µL and 600 µL. Vial size was 2-mL.

Maximizing the Ratio of Liquid to Headspace Volumes in the Vials

Intuitively, it might seem that with relatively more liquid phase, a significant enhancement of sensitivity is achieved. Using equation 2.5, polar compounds (where the partitioning favors the aqueous phase) exhibit virtually no increase in sensitivity when the ratio of the phases is altered. Table 2.6 demonstrates this with a sample of methanol in water. On the other hand, sensitivity is enhanced for non-polar samples when the proportion of liquid phase is increased. The magnitude of this effect varies with the partition coefficient.

Using Larger Sampling Vials

When sampling volatiles in water, larger sampling vials are not effective in enhancing sensitivity if the relative volumes of headspace and liquid are the same, as shown in Figure 2.8. Eisert and Pawliszyn [9] demonstrated that when sampling semivolatiles (pesticides) in water using fiber agitation, equilibrium was achieved more rapidly with 2-mL vials than with 16-mL vials. Therefore, a greater response was achieved in a shorter time with the smaller vials. The advantage of larger vials is that they are easier to fill with solid and semisolid samples; furthermore, with solids, larger aliquots are likely to be more representative of the original sample.

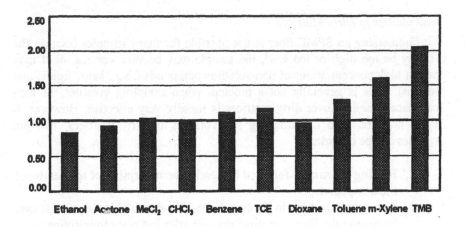

Figure 2.8 Relative responses with SPME sampling of the headspace over an aqueous mix in 16-mL vials versus 2-mL vials. In both cases, the liquid phase occupied 40% of the volume of the vial. (Courtesy of Varian Associates, Walnut Creek, CA 94598.)

Absorption and Desorption Times

When developing conditions for SPME sampling, one should plot detector response versus sampling time for each analyte in the sample. After the resulting curves are examined, the data will show that highly volatile analytes will reach a plateau in 15 minutes or less, indicating equilibrium. Compounds of lower volatility will show a steadily increasing response with time. For many volatile and semivolatile compounds, equilibrium is usually not achieved in a reasonable time (even with sample mixing).

Because maximum productivity is required in most laboratories, GC run times should be as short as possible; therefore, the SPME sampling time should be no longer than the total GC cycle time, minus the desorption time. Good precision can be achieved without attaining equilibrium if the absorption timing is precisely controlled [10].

For non-polar, volatile compounds, desorption is virtually complete in a few seconds, but the desorption should be continued for one or two minutes to ascertain that no carryover occurs when a blank is inserted after a sample. Some compounds, particularly polar semivolatiles, such as pesticides, exhibit significant sample carryover. This can usually be reduced to an acceptable level (> 0.5%) by selecting a fiber with a thin phase (e.g., 7 or 30 µm rather than 100 µm polydimethylsiloxane) and by desorbing for as long as 15-20 minutes.

Semivolatiles in Dirty Matrices

Immersing an SPME fiber is not possible for many samples because the pH may be too high or too low[4], the sample may be very viscous, or it may contain high concentrations of non-volatile compounds (e.g., salts, lipids, and proteins). This is generally not a problem when sampling volatiles; in fact, headspace sampling over dirty matrices is usually very effective. However, if SPME is to be used for sampling semivolatiles in dirty matrices, two approaches can be considered:

1. Heating the sample followed by headspace absorption of the semivolatile compound.
2. Immersing the fiber in the liquid phase to absorb the analyte, followed by rinsing the fiber to remove non-volatiles just prior to injection.

The first approach has been used for determining drugs and pesticides in body fluids [11-13]. Preliminary work with coffee beverages containing milk products indicated that heating was useful for increasing the response when sampling semivolatile flavor compounds such as vanilla (b.p. 285°C) (Figure 2.9). Note that heating was less useful for sampling benzaldehyde (b.p. 179°C).

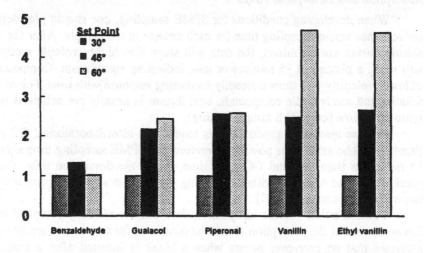

Figure 2.9 SPME sampling of the headspace over a flavored coffee sample at different temperatures. Heating the sample is more effective in increasing the response for polar compounds and compounds of low volatility. (Guiacol, vanillin, and ethyl vanillin are phenols.) (Courtesy of Varian Associates, Walnut Creek, CA 94598.)

---

[4] The pH range generally considered non-destructive to SPME fibers is between 2-11.

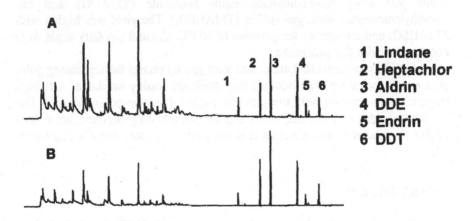

Figure 2.10 Whole milk was spiked with 100 ppb halogenated pesticides and the liquid phase was sampled with a 30-μm PDMS fiber. With chromatogram "A," the fiber was injected immediately after sampling; with "B," the fiber was washed in water for 6 seconds to remove the milk from the surface. (Courtesy of Varian Associates, Walnut Creek, CA 94598.)

Rinsing the fiber after liquid immersion appears to be effective in determining pesticides in whole milk. When the fiber was immersed in milk then injected into the hot injection port, a black, irregular coating appeared on the fiber after a few runs and the injector insert became contaminated. Figure 2.10 shows that if the fiber was rinsed with clean water for a few seconds after immersion in the milk, the pesticides were apparently retained on the fiber and the fat and protein were removed from the surface. This technique requires further study to verify that quantitation is unaffected by the rinsing.

Derivatization

Derivatization is used in gas chromatography to increase the volatility and improve the chromatographic performance of polar compounds. Pan and Pawliszyn [2] have derivatized fatty acids on SPME fibers. First, a mixture of C_{10}-C_{22} fatty acids in water was extracted with a PDMS fiber. After sufficient time had elapsed to extract the acids, the fiber was transferred to a second vial and exposed to the headspace over a diazomethane/diethyl ether solution. The derivatization reaction was complete in about 20 minutes.

The authors also investigated derivatization of the same acids in the injector port using tetramethylammonium hydroxide (TMAOH) and tetramethylammonium hydrogen sulfate (TMAHSO$_4$). The yield was higher with TMAHSO$_4$ and an injector temperature of 300°C. C$_{10}$ and C$_{12}$ fatty acids were not detected with this procedure.

Capillary columns that are coated with special phases for separating polar compounds including underivatized fatty acids are readily available. Although these columns perform well, they are less rugged than non-polar columns. Derivatization adds an extra step to the sample preparation procedure, but it does enable the analyst to use columns that are somewhat more durable and versatile.

QUANTITATION

For most samples, recovery of the analytes with SPME is relatively low; nevertheless, the degree of extraction is consistent so that SPME is a quantitative technique with excellent linearity, precision, and accuracy. The following discussion will describe liquid samples; solids will be described afterwards.

Liquid Samples

Matrix Effects

A major disadvantage of a sample preparation method that is based on equilibration rather than exhaustive extraction, is the necessity to consider matrix effects. Matrix effects are of major importance in quantitation with both static headspace and SPME. With these techniques, various components in a sample alter the partitioning between the phases. This is illustrated in Figure 2.11 where the recovery of various terpene alcohols was enhanced by salt and diminished by ethanol in the sample. As expected, various components of the matrix affect analytes differently, so that polar analytes are more affected by the presence of salt in the sample and non-polar analytes might be more affected by a fatty or oily matrix. "Standard additions" is a technique used for quantitation of samples where recovery is matrix-dependent.

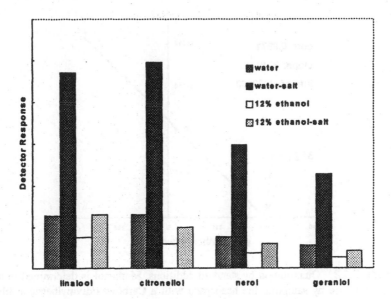

Figure 2.11 Illustrating the matrix effect on SPME recovery of terpene alcohols. Note that the "salting out" was very effective in enhancing the response in an aqueous solution, but the presence of ethanol reduced the effectiveness of the salt. (Courtesy of Varian Associates, Walnut Creek, CA 94598.)

Standard Additions

With many chemical analyses, calibration standards are prepared by adding known amounts of analytes to a blank that contains all of the components of the sample, except for the analytes of interest. The sample and the standard are analyzed, and the responses are compared to calculate the amount of unknown analyte in the sample.

If a blank matrix is not available, then known quantities of analyte are added to the unknown sample. The next step is the analysis of the unspiked sample plus the same sample spiked with at least two different levels. Next, a curve is plotted of detector response versus the quantity of analyte spiked into the sample. Normally, the curve is linear and a backward extrapolation will allow the analyst to calculate the quantity of analyte that was originally in the sample (Figure 2.12). The standard addition technique is commonly used not only with SPME and static headspace, but also with non-chromatographic techniques, such as atomic absorption spectroscopy.

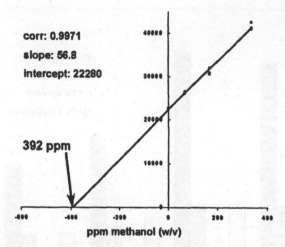

Figure 2.12 Quantitation by standard additions. Methanol is determined in a very caustic mixture by sampling the headspace with a Carbowax/divinylbenzene fiber. A blank matrix was not available so methanol was spiked into the sample at 100, 200, and 400-ppm levels. The resulting linear curve is back extrapolated to determine the initial methanol concentration. (Courtesy of Varian Associates, Walnut Creek, CA 94598.)

The standard addition technique will give accurate results, provided that the spiked analyte is recovered from the sample to the same degree as the analyte that is being measured. For example, drugs that are ingested may be bonded to proteins in biological fluids, and the same drug that is simply spiked into the sample may not be bonded. Hence, the recovery of the spiked drug will be greater than the drug that was ingested, giving a false low level. In this case, adding a reagent would be necessary to sever the bond between the ingested drug and the protein in order to use standard additions. Samples that are simple mixtures of water, salts, and small molecules should be amenable to standard addition quantitation.

Calibration by Comparison to a Standard

For samples where the matrix is available free of analytes, quantitation is achieved by spiking the blank matrix with the analytes to make calibration standards. After demonstrating that standards at different levels give a linear curve, a single calibration standard can be used for quantitation. Internal standards are often used in this type of calibration, but sometimes the sample contains several analytes and it is unlikely that a single internal standard will be suitable for tracking many analytes with different physical properties.

Sometimes samples in a fairly complex matrix, such as blood alcohol, can be analyzed using a single calibration standard if certain precautions are used.

When blood alcohol is determined with static headspace or with SPME, an internal standard is used, normally n-propanol. An internal standard is presumed to minimize matrix effects by exhibiting similar behavior to the analyte, thus eliminating the effect of slight differences among various samples and standards. However, the partition coefficient of ethanol was observed to be more sensitive to changes in salt concentration than the n-propanol (Z. Penton unpublished data).

Diluting the samples and standards, and saturating with salt can reduce subtle differences due to variations in organic compounds and ionic strength. Diluting the sample lowers the concentration of the analyte, but in many cases, including blood alcohol determination, sensitivity is still ample.

Preparation of Calibration Standards and Samples

Poor precision and accuracy often result from improper sample handling prior to SPME sampling. Some guidelines are listed below:

1. For standards and spiking solutions, a common practice is to prepare a stock solution of analytes in methanol, then add a small aliquot to water. This method is acceptable with SPME, and results are the same as adding organic compounds directly to water if the total level of methanol is less than 1%.
2. When adding standards to aqueous samples, the concentrations of all the components should be low enough so that they remain in solution.
3. Extra care is required for handling standards containing volatiles. When preparing standards of volatiles in water, the liquid should fill the entire storage container without any headspace. Caution should be exercised during procedures such as making serial dilutions, adding salt, adjusting the pH, and transferring from the container to the sampling vial. Aqueous standards and samples remaining in the containers that were used to fill the sampling vials should not be used again.

Solid Samples

Volatiles in solid samples can easily be detected with SPME, but accurate quantitation can be quite difficult. The analytes are often not distributed uniformly throughout the sample; for example, they may be concentrated on the surface or trapped within the interior. Moreover, a small portion of such a sample is unlikely to be representative. Sometimes, quantitation and even identification of analytes is unnecessary. SPME chromatograms of several small samples of packaging materials are shown in Figure 2.13. Various samples of a given material showed consistent patterns of organic contamination, and it was clear which material was the cleanest.

When quantitation is required, a solid can sometimes be dissolved in a small quantity of water or made into a slurry. This will help release organic impurities and increase the likelihood that the sample will be homogeneous so representative aliquots can be removed. The sample can then be analyzed by standard additions as described above or by spiking a blank matrix with a calibration standard.

Multiple Extraction

This technique was originally developed for the quantitative analysis of monomers in a polymer with static headspace [14], but it can also be applied to liquid samples and to SPME. With multiple extraction, a polymer is sealed in a vial and sampled repeatedly at equal time intervals. Under these conditions, the concentration of volatiles will decay exponentially, and if an infinite number of extractions are carried out, the volatiles will be completely removed from the vial. The total area count of the analyte is equal to the sum of the areas from each individual extraction.

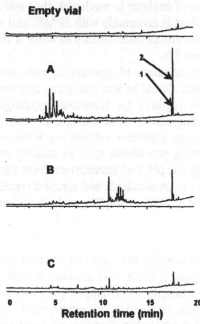

Figure 2.13 Packaging materials sampled with an SPME fiber. Each type of wrapping material gave a consistent pattern, allowing the manufacturer of a sensitive product to select the cleanest packaging. (Courtesy of Varian Associates, Walnut Creek, CA 94598.)

In practice, it is not necessary to extract more than three to six times, and using the following equation, the total area count can be calculated for each volatile in the sample:

$$\sum A_i = \frac{A_1}{1 - e^k}$$

Where:

$\sum A_i$ is the total area (2.6)

A_1 is the area count of the first extraction

k is the slope of the plot obtained by plotting the natural log of area counts versus the number of extraction steps. (k will be a negative number.)

The procedure is as follows:

1. Sample the polymer several times and determine the peak area *(A)* for each sampling.
2. Determine the natural log *(ln)* of *A*.
3. Plot *ln A* versus *n-1* where *n* is the number of samplings corresponding to *A*.
4. Determine the slope of the plot by linear regression.
5. Calculate the total area.
6. For calibration, prepare a vial that does not contain the matrix. The headspace volume in the vial should be equal to the headspace volume in the sample vials. (The calibration vials can be filled with glass beads with a volume that is the same as the volume of the samples.)
7. Inject a known quantity of the analyte of interest into the calibration vial.
8. Following steps 1-5, calculate the area corresponding to the known standard.
9. Calculate the amount of volatile in the unknown, by comparing the area of the calibration standard to the area of the unknown (external standard calculation).

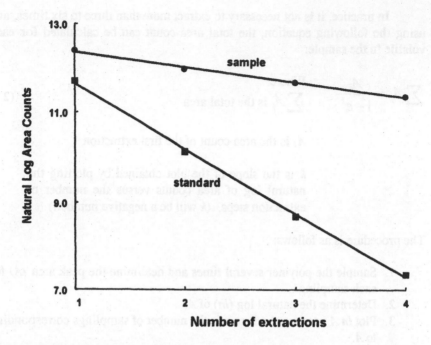

Figure 2.14 Plot of multiple extractions of vinyl chloride from a polyvinyl chloride polymer and from a standard. The matrix effect resulted in a slower rate of extraction of the monomer from the polymer than from the vial containing the calibration standard. (Courtesy of Varian Associates, Walnut Creek, CA 94598.)

Figure 2.14 shows a graph for a calibration standard and an unknown in the analysis of vinyl chloride in polyvinylchloride. After demonstrating a linear response to validate the method, such as that shown in the graph, a simplified form of equation 2.6 can be used, which requires only two samplings:

$$\sum A_i = \frac{A_1^2}{(A_1 - A_2)}$$

Where:

$\sum A_i$ is the total area count

A_1 is the area count of the first extraction

A_2 is the area count of the second extraction.

(2.7)

While results would be expected to be more accurate with more than two samplings, equation 2.7 is practical for routine analysis. The area counts in the calibration would also be determined using equation 2.7.

Generally, when a new quantitative method is being developed using SPME or another technique, the sample should be analyzed by the new method and by a quite different technique, then the results compared. If the results are the same with both methods, the analyst can be confident that the new method is valid; if results are different, the reason for the discrepancy should be investigated.

TROUBLESHOOTING

Although the popularity and increasing use of SPME is partially due to the simplicity of the technique, the new user should be aware of possible problems and how to approach them. Table 2.7 lists some of the difficulties that may be encountered.

Table 2.7 Troubleshooting SPME.

Problem	Possible Causes and Solutions
No peaks.	Verify that the fiber is still attached to the rod and that the coating hasn't stripped off (refer to Chapter 3).
	Is the column head pressure unchanged and does the pressure stay constant during injection?
	In splitless injection, is the splitter closed for at least 2-3 minutes?
	If there is no obvious problem with the fiber or the septum, make a liquid injection with a standard containing the analytes in a suitable organic solvent to determine if the GC system has a problem.
Extraneous peaks persist in blanks after an initial morning "bakeout" run is made to remove contaminants in the fiber.	Has the fiber been conditioned properly?
	Program the column without making an injection to verify that the GC system is not contaminated.
	Sample an empty vial and if the extraneous peaks are still present, bake the vial septa for several hours at 150°C.
Fiber breakage, fiber sheath and/or rod bent.	The injector insert may be plugged with septum particles. (See section on preparation of the GC for SPME.)
	If observed with automation, observe the sampling and desorption steps. Consult the autosampler manual for information on aligning the autosampler.

Table 2.7 Continued.

Problem	Possible Causes and Solutions
Sudden loss of sensitivity.	The fiber has broken or some of the coating has been stripped. (Occasionally fibers are defective.)
	Injector septum is leaking.
Poor precision (RSD's should generally be 1-10%).	Manual SPME: the timing of absorption and desorption may not be precise.
	Manual SPME: the position and depth of the fiber in the injector insert may vary.
	With volatile analytes, the problem may be poor sample handling so that analytes are lost during transfer to the sample vial.
	Were equal volumes of samples added to the vials? Inaccurate pipetting will contribute to poor precision, especially for non-polar compounds.
	SPME fiber may be aged.
	Also may be a GC problem—activity in the column, injector insert.
Carryover.	Try a longer desorption time and/or higher injector temperature.
	Examine the fiber to determine if the coating is intact.
	With liquid sampling, the level of the liquid in the sample vial should not be above the top of the fiber. The metal rod should be extended slightly past the sheath during sampling.
	The problem may be due to the injector or the column. Examine the insert to confirm that it is clean. Make injections with standards in an organic solvent and see if the carryover is due to absorption in the column.
Broad peaks early in the run.	Incorrect injector insert.
	Initial column temperature too high to focus volatile compounds.

REFERENCES

1. L Pan, M Adams, J Pawliszyn. Determination of fatty acids using solid-phase microextraction. Anal Chem 67:4396-4403, 1995.
2. L Pan, J Pawliszyn. Derivatization/solid-phase microextraction: new approach to polar analytes. Anal Chem 69:196-205, 1997.
3. JL Snyder. The use of gas chromatography and headspace sampling in environmental analysis presented at Eastern Analytical Symposium, Somerset, NJ, November, 1997.
4. NH Snow, P Okeyo. Initial bandwidth resulting from splitless and Solid Phase Microextraction gas chromatographic injections. J High Resolut Chromatogr 20:77-80, 1997.
5. Z Zhang, J Pawliszyn. Headspace solid-phase microextraction. Anal Chem 65:1843-52, 1993.
6. CL Arthur, LM Killam, S Motlagh, M Lim, DW Potter, J Pawliszyn. Analysis of substituted benzene compounds in groundwater using solid-phase microextraction. Environ Sci Technol 26:979-983, 1992.
7. BV Ioffe, AG Vitenberg. Headspace Analysis and Related Methods in Gas Chromatography. New York: John Wiley & Sons, 1984, pp 61-62.
8. Z Zhang, J Pawliszyn. Quantitative extraction using an internally cooled Solid Phase Microextraction device. Anal Chem 67:34-43. 1995.
9. R Eisert, J Pawliszyn. Design of automated Solid Phase Microextraction for trace analysis of organic compounds in aqueous samples. J Chromatogr A 776: 293-303, 1997.
10. J Ai. Headspace Solid Phase Microextraction. Dynamics and quantitative analysis before reaching a partition equilibrium. Anal Chem 69:3260-3266, 1997.
11. T Kumazawa, X Lee, M Tsai, H Seno, A Ishii, K Sato. Simple extraction of tricyclic antidepressants in human urine by headspace Solid Phase Microextraction. Jpn J Forensic Toxicol 13: 25-30, 1995.
12. T Kumazawa, K Watanabe, K Sato, H Seno, A Ishii, O Suzuki. Detection of cocaine in human urine by Solid Phase Microextraction and capillary gas chromatography with nitrogen-phosphorous detection. Jpn J Forensic Toxicol 13:207-210, 1995.
13. X-P Lee, T Kumazawa, K Sato, O Suzuki. Detection of organophosphate pesticides in human body fluids by headspace Solid Phase Microextraction and capillary gas chromatography with nitrogen-phosphorous detection. Chromatographia 42:135-140, 1996.
14. B Kolb. Multiple headspace extraction—a procedure for eliminating the influence of the sample matrix in quantitative headspace gas chromatography. Chromatographia 15:587-594, 1982.

3

SPME Fibers and Selection for Specific Applications

Robert E. Shirey
Supelco, Bellefonte, Pennsylvania

FIBER ASSEMBLIES AND HOLDERS

The fiber assembly and holder design must provide maintenance of sample integrity, ease of use, and versatile fiber usage. Most importantly, the assembly must be leak tight to prevent sample loss during the extraction and desorption steps. Secondly, the holder must easily expose and retract the fiber and be similar to using a typical liquid syringe. Likewise, the fiber assembly needle must be of sufficient diameter to contain a fiber, yet small enough to puncture a septum without coring. Finally, easily replacing fiber assemblies in the holder is essential.

Figure 3.1 shows a detailed view of the manual fiber assembly. It contains two pieces of tubing: an outer piece of 24-gauge tubing that performs as a needle and an inner piece of 26-gauge tubing that contains the fiber. The fiber is inserted and glued into the inner tubing with a high-temperature, solvent-resistant epoxy. The top of the outer tubing is sealed with a septum contained on a ferrule that prevents carrier gas from leaking through the outer needle. The inner tubing is pierced through the septum and sealed at the top in a color-coded hub, which identifies the type of fiber attached. The hub contains threads that enable the fiber to be easily screwed in and out of the fiber assembly holder, thereby, providing ease in changing fibers.

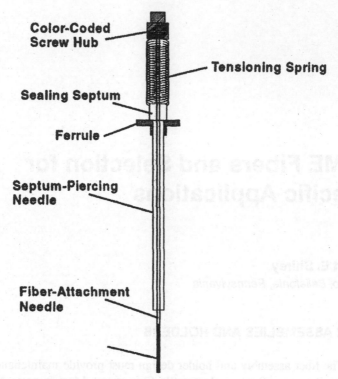

Color-Coded
Screw Hub

Tensioning Spring

Sealing Septum

Ferrule

Septum-Piercing
Needle

Fiber-Attachment
Needle

Figure 3.1 SPME fiber assembly for manual holder.

Two styles of fiber assemblies are commercially available: one for use in the manual holder and the other for use in a holder for the Varian 8200 Auto-Sampler and for the HPLC interface (Figure 3.2). The primary difference between the two is that the manual version contains a spring that keeps the fiber retracted when not in use, but this is not necessary in the automated version. Manual fiber assemblies cannot be used with the autosampler or HPLC interface. However, the fiber assemblies without springs can be used with a manual holder, if the plunger is physically pulled back to retract the fiber. The manual version also has a "Z"slot to lock the fiber in the exposed position for sampling and desorption; moreover, it provides a depth gauge and needle guide to control the depth that the fiber goes into the injection port or sampling vial.

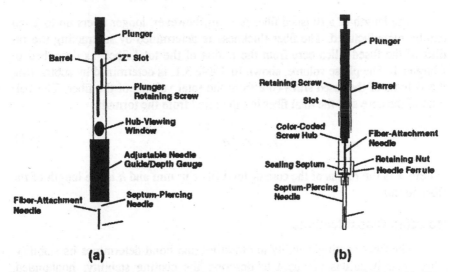

Figure 3.2 SPME fiber assembly holders: (a) manual fiber assembly holder, (b) fiber assembly holder for autosamplers or HPLC interface.

FIBER COATING PREPARATION AND TYPES

Fiber Preparation

All manufactured SPME fibers contain a fused silica (FS) core. The diameter of the fused silica is 110μm for most of the fibers; although, the diameter of the pre-coated core used with HPLC fibers is 160 μm. To make the single component fiber coatings, i.e., poly(dimethylsiloxane) (PDMS) and polyacrylate (PA), the fiber is first drawn in a tower. Then, the fiber is passed through a specified diameter orifice in a cup that contains the desired phase that coats the fiber. This process is accomplished in one step, producing about 1,000 m of coated fiber. Consequently, 50 to 75 thousand fiber assemblies are manufactured at one time, which assures that little variation among fibers occurs. Finally, each lot of fiber is tested and compared to previous fiber production lots.

For multiple component coatings, such as divinylbenzene (DVB) and Carboxen™ suspended in PDMS or Carbowax®, the fibers must be hand coated in multiple coating steps. The fiber coating is on 110 μm fused silica, then each fiber is measured to assure consistency. The hand coating results in slightly greater variation, but the fibers are tested to assure reproducibility.

The length of a finished fiber is 1 cm; however, longer fibers up to 2 cm can be manufactured. The fiber thickness is determined by subtracting the radius of the fused silica core from the radius of the total fiber, as described in Chapter 1. The phase volume, shown in Table 3.1, is determined by subtracting the volume of the fused silica core from the total volume of the fiber. The volume of the core and the total fiber is calculated from the formula:

$$V = \pi r^2 h \tag{3.1}$$

Where: r is the radius of the core or total fiber in mm and h is the length of the fiber in mm.

Bonding Classifications

The fiber coating's ability to crosslink and bond determines its stability. Three classifications are used to describe the coating stability: nonbonded, crosslinked, and bonded. Nonbonded phases are stabilized, but do not contain any crosslinking agents. These phases are not solvent resistant and tend to swell in organic solvents; however, we have determined that they can withstand some polar organic solvents, such as methanol and acetonitrile. Finally, nonbonded fiber coatings have less thermal stability than bonded fiber coatings.

Table 3.1 Phase Volume of Commercially Available SPME Fibers.

Fiber Coating	FS Core Diameter (mm)	FS Core Radius (mm)	Core Vol. (mm^3)	Total Diameter (mm)	Total Vol. (mm^3)	Phase Vol.(mm^3 or μl)
PDMS, 100 μm	0.110	0.055	0.095	0.300	0.707	0.612
PDMS, 30 μm	0.110	0.055	0.095	0.170	0.227	0.132
PDMS, 7 μm	0.110	0.055	0.095	0.124	0.121	0.026
Polyacrylate, 85 μm	0.110	0.055	0.095	0.280	0.616	0.521
CW-DVB, 65 μm	0.110	0.055	0.095	0.240	0.452	0.357
PDMS-DVB, 65 μm	0.110	0.055	0.095	0.240	0.452	0.357
Carboxen-PDMS, 75 μm	0.110	0.055	0.095	0.260	0.531	0.436
CW-TPR HPLC, 50 μm	0.160	0.080	0.201	0.260	0.531	0.330
PDMS-DVB HPLC, 60 μm	0.160	0.080	0.201	0.280	0.616	0.415

Partially crosslinked fiber coatings contain crosslinking agents, such as vinyl groups. The phase crosslinks with itself producing a more stable coating; however, the coating does not tend to bond to the fused silica. Partially crosslinked fiber coatings are more solvent resistant than nonbonded coatings and have better thermal stability. Most of these fibers can be exposed to a variety of solvents, but special care must be taken to prevent the fiber coating from being stripped off the fused silica core due to swelling.

Bonded fiber coatings, like partially crosslinked coatings, contain crosslinking agents. The difference is that the coating is crosslinked not only to itself, but also to the fused silica. These coatings are very resistant to most organic solvents and have good thermal stability; furthermore, they can be rinsed in organic solvent with minimal swelling. These coatings tend to be thinner.

The thicker the coating, the more difficult it is to truly bond; consequently, only one bonded phase, 7 µm PDMS, is currently available. Table 3.2 outlines the commercially available fibers, bonding type, temperature limitations, hub color (for identification), and phase polarity.

The fibers manufactured specifically for HPLC are coated over a specially pre-coated fused silica fiber that reduces the chance of breaking. The coating appears to partially bond to the pre-coating; moreover, the pre-coating has little or no effect on the extraction of the analytes. The thickness of the fused silica core and pre-coating is 160 µm. Because the pre-coating is pliable, the fiber can be crimped in the tubing rather than glued, which eliminates UV observed peaks that would otherwise be obtained from the glue.

Table 3.2 Commercially Available SPME Fibers.

Fiber Coating Type	Coating Stability	Max Temp	Hub Color	Polarity
PDMS ,100 µm	nonbonded	280°C	red	nonpolar
PDMS, 30 µm	nonbonded	280°C	yellow	nonpolar
PDMS, 7 µm	bonded	340°C	green	nonpolar
Polyacrylate, 85 µm	crosslinked	320°C	white	polar
CW-DVB, 65 µm	crosslinked	260°C	orange	polar
PDMS-DVB, 65 µm	crosslinked	270°C	blue	bipolar
Carboxen-PDMS, 75 µm	crosslinked	340°C	black	bipolar
CW-TPR HPLC, 60 µm	crosslinked	—	brown-notched	polar
PDMS-DVB HPLC, 50 µm	crosslinked	—	purple-notched	bipolar

Fiber Coatings and Properties

Liquid Phases – Poly(dimethylsiloxane), Polyacrylate, and Carbowax

Fiber coatings are generally classified by polarity and film thickness. The most common nonpolar phase is poly(dimethylsiloxane), which is similar to OV®-1 and SE-30 type GC phases. As shown in Table 3.1, three PDMS fibers are available with film thicknesses of 100 μm, 30 μm and 7 μm. A thicker coating extracts more of a given analyte, but the extraction time is longer than for a thinner coating.

The more polar phases are polyacrylate and Carbowax (CW). Polyacrylate is a rigid material that is not a liquid at room temperature, in contrast to most phases. Because it is rigid, migration of the analytes in and out of the fiber coating is slower. Thus, it takes longer to extract analytes with the polyacrylate fiber than with other conventional fibers. Additionally, higher temperatures are required to desorb the analytes, and early eluting peaks may tail if not properly focused at the head of the GC column. Polyacrylate fibers are quite solvent resistant and durable; however, the fibers can be oxidized easily at high temperatures. Since higher injection port temperatures (>250°C) are required, the carrier gas must be free of oxygen. The polyacrylate fiber will darken to dark brown when exposed to temperatures greater than 280°C, but this darkening process is normal and does not adversely affect the fiber. However, if the fiber turns charcoal black, the fiber has been oxidized and is most likely not suitable for further use.

Carbowax has been used as a moderately polar phase in gas chromatography; therefore, it is a natural choice for use as an SPME phase coating. Nevertheless, it has some drawbacks that need to be addressed. First, Carbowax tends to swell or dissolve in water. Moreover, swelling of the phase coating cannot be tolerated because it risks being stripped off the fiber when the inner needle is retracted into the outer needle. To overcome this problem, a highly crosslinked Carbowax was synthesized. The high crosslinking reduces swelling and solubility in water, but also slightly reduces the polarity. Secondly, Carbowax, like polyacrylate, is sensitive to oxygen at temperatures above 220°C. Furthermore, oxygen in the carrier gas or leaking through the septum can oxidize the fiber coating. When oxidation occurs, the fiber darkens, the coating becomes powdery, and it can be easily removed from the fused silica. Several precautions can be taken to reduce this problem:

1. Use appropriate carrier gas purifiers or 99.999% helium. Heated catalytic purifiers are best for helium, and oxygen indicator tubes containing alkyl lithium or MnO are suitable for hydrogen carrier gas; however, carrier gas purifiers made from plastics are not appropriate.

Fiber life appears to be shorter when hydrogen is used as a carrier because apparently, it is not as pure as helium.

2. Using an injector temperature range of 180-240°C will lengthen fiber life. If the carrier gas is oxygen free, the fiber can withstand 260°C.
3. Fibers should not be conditioned for longer than 1 hour at temperatures above 220°C.
4. Desorption times should be limited to 5 minutes.

Previously, a limitation in producing Carbowax fibers was the ability to make thick coatings (>30 μm). To overcome this problem, Carbowax was blended with divinylbenzene, a porous polymer that has pore properties similar to HayeSep® Q and Porpak® Q, which is discussed next.

Porous Particle Blends

Divinylbenzene
Pores have the ability to adsorb analytes and physically retain them, which results in tighter retention of the analytes that fit into the pores. Therefore, fibers containing porous materials are generally better for trace level analysis (ppt or ppb) or for ppm levels of analytes with low distribution constants.

The surface area of divinylbenzene (DVB) is approximately 750 m²/g; moreover, the porosity distribution is primarily mesoporous, with some macro and micro pores. Refer to Table 3.3 for pore size and distribution. The micropores of DVB are fairly large with an average diameter of 17 Å. The mesopores are ideal for trapping C_6-C_{15} analytes, but DVB can also trap and release larger molecules efficiently. DVB has a high degree of porosity at 1.5 mL/g.

Because DVB is a solid particle, it must be suspended in a liquid phase to coat it onto the fiber. Accordingly, this blending provides advantages from both phases. Blending DVB with Carbowax increases the polarity of the fiber and its ability to extract a wide molecular weight range of analytes. DVB can also be blended with PDMS. This combination provides slightly better retention of smaller analytes than PDMS alone, and it has been shown to have affinity for some polar analytes. Applications using both DVB blends will be discussed in the application section later in this chapter. The limitation of DVB fibers is that the coatings are more fragile and can be stripped off the fibers. Consequently, extra precautions are needed to protect the fibers. At the time of this writing, research to better bond the coating to the fused silica core has been quite encouraging. The addition of a crosslinking catalyst in the phase has greatly improved the stability of the coating.

Table 3.3 Physical Properties of Divinylbenzene and Carboxen 1006.

Material	Surface Area		Porosity mL/g		
	m²/g	Macro	Meso	Micro	Total
Divinylbenzene	750	0.58	0.85	0.11	1.54
Carboxen 1006	720	0.23	0.26	0.29	0.78
	macropore ≥500 Å, mesopore = 20-500 Å, micropore = 2-20 Å				

Carboxen 1006

Carboxen 1006 has several favorable characteristics that, when mixed with PDMS, create a bipolar phase for smaller analytes. These unique characteristics include pore size, distribution, volume, shape, and particle size. First, Carboxen is a line of porous, synthetic carbons that has distinctive pore designs; specifically, Carboxen 1006 has an even distribution of micro, meso, and macro pores as shown in Table 3.3. Pore size is critical in determining which analytes are more tightly retained. For optimum extraction, the pore diameter should be about twice the size of the molecule that is being extracted [1]. The average micropore diameter is slightly larger than 10Å with a distribution from less than 2 to 20 Å, which is ideal for trapping small molecules. Nevertheless, Carboxen 1006 does not retain extremely small analytes, such as methane, nitrogen, and oxygen, but ethane is slightly retained. Moreover, it has a high percentage of porosity, as indicated by the total pore volume of 0.78 mL/g, which gives it more trapping surface area. The surface area of Carboxen 1006 is 720 m²/g, when measured with a single point reference; furthermore, when multiple reference points are used to measure surface area, it exceeds 950 m²/g.

Second, because Carboxen 1006 is synthetically produced, the particle size can be carefully controlled in addition to pore volume and size. Microscopic analysis of Carboxen particles in SPME fibers reveals a distribution between 1 to 4 μm, with an average size of 2 μm. These small particles enable multiple layers to be coated on the fiber, which increases the total fiber capacity.

Third, not only must the analytes be adsorbed into the pore, but they must also be efficiently desorbed. The shape that the pore takes can affect the rate of adsorption and desorption. Carboxens have a unique feature compared to most other porous carbons, that is, the pores travel through the entire length of the particle. This characteristic, known as throughput, is shown in the Carboxen particle schematic in Figure 3.3. The throughput of the pores enables small analytes to be more rapidly desorbed because they can pass through the pore.

On the other hand, pores that do not have throughput will result in carryover or extremely slow desorption of the analytes. Usually, poor desorption is caused by hysteresis, i.e., the condensation of an analyte in the pore. Typically, hysteresis occurs with midsize volatiles, such as toluene and xylenes, that are trapped in the mesopores and must reverse direction to be desorbed. Additionally, closed pores or slit-like pores have a much higher degree of hysteresis than the open tapered pores of Carboxen [1]. High desorption temperatures help to minimize the effects of hysteresis by releasing moderately large molecules that are trapped in the macropores. For example, increasing the desorption temperature to greater than 280°C can reduce the amount of peak tailing. Consequently, applications with this fiber will often recommend desorption temperatures of 310° to 320°C.

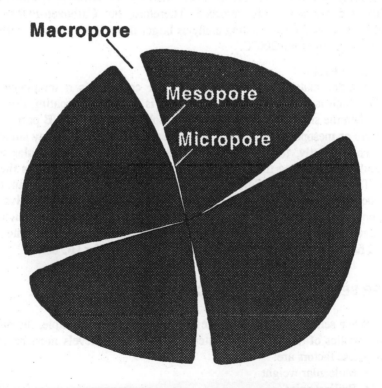

Figure 3.3 Schematic view of the pore structure in Carboxen 1006.

Carbosieves, another line of synthetic porous carbon, and other Carboxens with smaller pore diameters have also been investigated for SPME use. However, the rates of adsorption and desorption for the totally microporous carbons were much slower than for Carboxen 1006 [2]. Furthermore, many of the analytes were not released from the fibers, and even elevating temperatures to 360°C did not sufficiently remove analytes from Carbosieve S-III. Hence, these fiber blends do not appear promising at this time.

In conclusion, the combination of pore distribution, volume, and surface area and particle shape makes Carboxen 1006 a good material for SPME fibers, as will be shown in the application section later in this chapter. It has sufficient pore variety and surface area to extract and efficiently desorb molecules in the C_2-C_{12} range. However, molecules larger than C_{12} are strongly retained on its surface and are not easily desorbed. Therefore, the Carboxen-PDMS fiber should not be used for extracting analytes larger than C_{12} or analytes with boiling points greater than 200°C.

Templated Resin

Another material used for SPME fiber coatings is a templated resin (TPR), which is a hollow spherical DVB. It is initiated by coating DVB over silica, then the silica is dissolved leaving a hollow spherical DVB particle with no micro or mesopores. The result is a material that excludes many small analytes; consequently, the templated resin can extract higher molecular weight molecules. Suspending the templated resin in Carbowax then makes the final fiber. The polar nature of Carbowax, combined with the templated resin, makes the fiber suitable for extracting large polar molecules. Moreover, a solvent can efficiently remove analytes, which makes the fiber ideal for HPLC applications. Like the PDMS-DVB HPLC fiber, the CW-TPR material is coated onto a pre-coated fused silica fiber. This fiber is not recommended for GC use.

FIBER SELECTION AND APPLICATIONS

When selecting the appropriate fiber for extracting a sample, the physical characteristics of the analytes of interest and detection levels must be considered. These factors are:
- Molecular weight
- Boiling point
- Vapor pressure
- Polarity
- Functional group
- Concentration range
- Detector type

These factors need to be matched with the variables in fiber coatings, such as film thickness, polarity and porosity. This section describes how to match the appropriate fiber coating to the analyte characteristics. Near the end of this section, a fiber selection guide is presented to help you select the appropriate fiber by analyte class. Note that you may also need to modify the sample to optimize extraction efficiency, as covered in Chapter 2.

Effects of Fiber Coating Thickness

Recall from Chapter 1 that a thick fiber coating will extract more of a given analyte than a thin coating. A larger phase volume can retain more analyte; consequently, it can yield a broader linear range. Nevertheless, if sampling time is critical, thinner fiber coatings may be more advantageous. Analytes with high distribution constants tend to be larger and more slowly adsorbed. Because their migration into and out of the fiber is slow, a thinner film can retain a large portion of these analytes. If the distribution constant is high, there is a high affinity for the fiber so a thin film is adequate.

For retention of volatile compounds with low molecular weight, a thicker film is advantageous to adequately retain the volatile analytes. An example of the differences in film thickness is shown in Table 3.4, where the extraction efficiency of three PDMS fibers is compared for a wide range of aromatics. The 16 aromatics were extracted by direct immersion for 15 minutes. Chrysene was chosen as the internal standard because its responses were nearly identical for all three fibers. The relative recoveries show the effect of molecular size for the polyaromatic hydrocarbons (PAHs). Larger molecules, such as benzo(g,h,i)-perylene, move more slowly compared to the smaller analytes. The thick coating of the 100 μm fiber retains more of the volatile analytes (i.e., benzene and toluene) with respect to the other fibers, but does more poorly for the larger PAHs in the given time period. If the extraction time is longer, those values will increase. The 7 μm PDMS fiber extracts the larger PAHs most efficiently, but poorly retains the smaller PAHs. The larger analytes can migrate more readily into a thinner coating. Hence, the 30 μm PDMS fiber is a good compromise. It can retain enough of the smaller analytes for detection, but can also extract the larger analytes more efficiently than the 100μm PDMS fiber.

The 30 μm PDMS fiber is ideal for characterizing underground tank leaks where sample analysis includes differentiating between diesel fuel and gasoline and quantifying BTEX and other petroleum volatile organic compounds (PVOC) in the sample. Figure 3.4 shows the extraction of both gasoline and diesel fuels by the same fiber using identical extraction conditions. The 30 μm PDMS fiber easily extracts many of the light volatiles in gasoline as well as the high boiling components in diesel fuel. With use of GC and SPME, the types of fuels leaking from underground tanks can be easily identified.

Table 3.4 The Relation of Phase Coating Thickness to Molecular Size as Measured by Percent Relative Recovery.

Analyte	No. of Rings	% Relative Recovery for each PDMS Film Thickness		
		100μm	30μm	7μm
Benzene	1	2	1	<1
Toluene	1	5	1	<1
Ethylbenzene	1	6	4	1
1,3-Dichlorobenzene	1	15	5	1
Naphthalene	2	13	4	1
Acenaphthylene	3	19	8	3
Fluorene	3	29	18	6
Phenanthrene	4	37	27	16
Anthracene	4	49	38	32
Pyrene	4	69	54	47
Benzo(a)anthracene	5	105	91	96
Chrysene	5	100	100	100
Benzo(b)fluoranthene	5	104	111	120
Benzo(a)pyrene	5	119	127	131
Indeno(1,2,3-cd)pyrene	6	61	140	148
Benzo(g,h,i)perylene	6	61	117	122

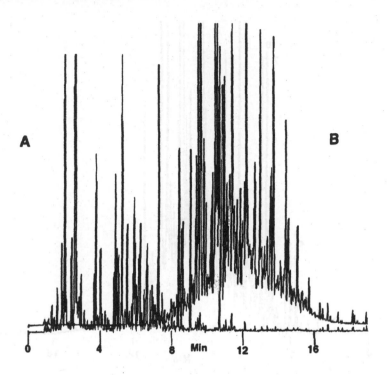

Figure 3.4 Analyses of gasoline and diesel fuel extracted from water by SPME. Gasoline (A) at 2 ppm in water and diesel (B) at 1 ppm were extracted for 15 min by direct immersion using 30 μm PDMS fiber. The analysis was by GC/FID, using a 15 m x 0.20 mm x 0.20 μm SPB™-1 column. Oven temperature was 40°C (hold 1 min) to 320°C at 15°C/min. Injection port temperature was 250°C. A splitless 0.75 mm ID liner was used. The split vent was closed for the entire desorption time of 4 min.

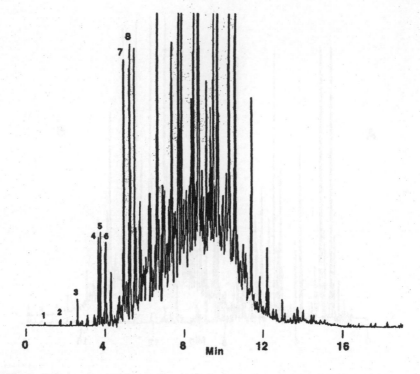

Figure 3.5 Analysis of PVOCs and kerosene from water by SPME: 1. MTBE; 2. benzene; 3. toluene; 4. ethylbenzene; 5. m-xylene; 6. o-xylene; 7. 1,3,5-trimethylbenzene; and 8. 1,2,4-trimethylbenzene. These components were spiked at 50 ppb in water with 4 ppm of kerosene. The extraction and analytical conditions were the same as in Figure 3.4.

Figure 3.5 shows the ability to extract trace amounts of PVOCs that are commonly monitored in the presence of a large amount of kerosene. This demonstrates that the fibers can extract analytes in low concentrations in the presence of other analytes in high concentrations. All of the fuel extractions were accomplished by directly immersing the fiber for 15 minutes.

Film thickness can play a major role in extraction precision due to carryover and lengthy extraction times, especially for analytes with high distribution constants. These analytes are usually larger, nonpolar analytes, such as chlorinated pesticides, PAHs, and long chain hydrocarbons. Reaching equilibrium may take hours for some of the analytes when using the 100μm PDMS fiber. Approximately 30 minutes, or the cycle time for one chromatographic analysis, is usually the practical limitation for setting an extraction time. Because this time is not necessarily at equilibrium, slight changes in variables such as time, temperature, stirring rate, and organic content in the water sample, can dramatically change the results, which can lead to poor reproducibility.

Poor precision can also be caused by incomplete desorption of these large analytes. We have determined that carryover generally is less than 0.1%; although, the carryover is larger and more random with some of these analytes. Increasing the injection port temperature may help, but this is not always effective.

The use of a thinner fiber coating can help to improve the precision. Reducing PDMS film thickness from 100 μm to 30 μm can greatly reduce the time needed to reach equilibrium without a significant loss in linear range. By going to an even thinner fiber coating, such as the 7 μm PDMS, the time to reach equilibrium is less, but the limited capacity of the fiber reduces linear range. Several applications, such as PAHs and chlorinated pesticides, were initially developed before the 30 μm fiber was introduced. Since these initial applications, the 30 μm PDMS fiber may be a better choice than either the 7 μm or 100 μm PDMS fiber.

Carryover is somewhat reduced with thinner fiber coatings because these coatings retain less analyte and can be desorbed at higher temperatures than thicker coatings of the same phase. These factors do not significantly reduce the actual amount of carryover, but the relative amount is held more closely to the 0.1% level, which helps to improve the precision of the extractions. It is surprising that a thin coating of a liquid phase could retain any of the analytes when exposed to high injection port temperatures. Perhaps some of the carryover is due to interactions with the fused silica core.

To reach equilibrium, the analytes must migrate through the entire radius of the liquid phase coating to contact the fused silica core. Even though fused silica is inert, some analytes, such as chlorinated pesticides, can interact with fused silica, which could result in carryover of some types of analytes. Deactivation of the fused silica core has been investigated, but the results are inconclusive.

Analyte desorption from a liquid phase is generally rapid as shown in the fast analysis of VOCs in Figure 3.6. In this example, the VOCs are rapidly desorbed from the 100μm PDMS fiber onto two short, narrow bore columns.

Because no solvent is used with SPME, the analytes can be rapidly focused on the head of the column. Moreover, with only nanogram amounts of analytes being desorbed, narrow bore columns can be utilized. These columns have more theoretical plates per meter than wider bore columns; consequently, shorter columns may be used to sharpen peaks and significantly reduce analysis time. Dual columns were used to provide a conformational screening method that could be used with an FID. Samples ranging from 25 ppb to 10 ppm have been evaluated with good linearity, as shown in Table 3.5.

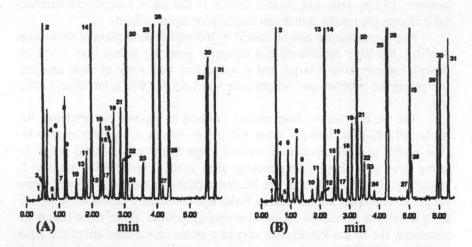

Figure 3.6 Screening analysis of VOCs at 25 ppb in water. SPME extraction split into two capillary columns installed with a graphite ferrule. **(A)** SPB-1 10 m x 0.20 mm x 1.2 μm column and **(B)** VOCOL™ Column of same dimensions. The analytes are listed in Table 3.6. Extraction (direct immersion) was for 5 minutes with 100μm PDMS fiber. The oven program was 40°C (0.75min) to 180°C at 20°C/min. The split/splitless injection port temperature was set at 240°C and a 0.75 mm ID liner was used. The vent was closed for the entire desorption time of 3 min.

Screening of VOCs is performed at Lancaster Laboratories in Lancaster, PA. This laboratory's method includes a 30 μm PDMS fiber and a 12-second headspace extraction to monitor commonly encountered VOCs that are present in high concentrations. A 30 μm fiber is selected because it more rapidly extracted and desorbed the analytes. By screening samples that may contain high concentrations of VOCs, the samples can be diluted prior to Purge and Trap/GC/MS analysis. This saves down time and greatly reduces the need to reanalyze samples because Purge and Trap analysis has an upper level concentration range of 200 ppb [3]. Samples with higher concentrations will be out of the calibration range and can also severely contaminate the purge trap.

Table 3.5 Response Factors[a] Calculated Over Nine Concentration Levels.

Analyte	% RSD	Analyte	% RSD
1. Chloromethane	17.6	17. Bromodichloromethane	8.9
2. Vinyl chloride	21.0	18. 2-Chloroethylvinyl ether	6.7
3. Bromomethane	11.8	19. cis-1,3-Dichloropropene	4.1
4. Chloroethane	16.8	20. Toluene	8.8
5. Trichlorofluoromethane	8.7	21. trans-1,3-Dichloropropene	4.7
6. 1,1-Dichloroethene	12.1	22. 1,1,2-Trichloroethane	4.2
7. Methylene chloride	15.8	23. Tetrachloroethene	20.3
8. trans-1,2-Dichloroethene	14.8	24. Dibromochloromethane	6.2
9. 1,1-Dichloroethane	8.8	25. Chlorobenzene	9.8
10. Chloroform	11.1	26. Ethylbenzene	24.8
11. 1,1,1-Trichloroethane	4.7	27. Bromoform	6.6
12. Carbon tetrachloride	8.6	IS 1,4-Dichlorbutane	
13. Benzene	4.7	28. 1,1,2,2-Tetrachloroethane	6.0
14. 1,2-Dichloroethane	3.7	29. 1,3- Dichlorobenzene	28.1
15. Trichloroethene	4.0	30. 1,4 -Dichlorobenzene	27.0
16. 1,2-Dichloropropane	8.9	31. 1,2-Dichlorobenzene	27.2

$$\text{[a] } Response\ Factor = \frac{\left(area\ counts_{analyte}\right) \times \left(concentration_{int.\ std.}\right)}{\left(area\ counts_{int.\ std.}\right) \times \left(concentration_{analyte}\right)}$$

Figure 3.7 shows a typical chromatogram of the screening analysis. The correlation between Purge and Trap and SPME was excellent; moreover, the mean percent agreement range of the individual analytes was 90-115%. Total cycle time was critical in the development of a screening method. With use of a Varian autosampler, the cycle time per sample was less than four minutes. Twenty-five percent salt (NaCl) was needed to enhance sensitivity and precision.

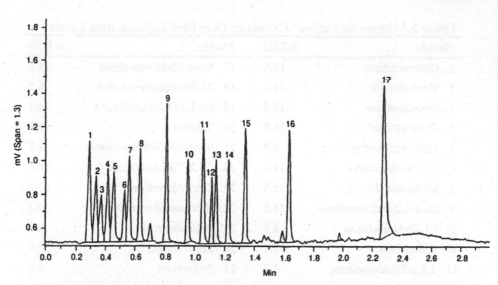

Figure 3.7 Rapid screening analysis of commonly encountered VOCs usually present at high concentration levels. The analytes are: 1. methanol; 2. acetone; 3. methylene chloride; 4. MTBE; 5. cis-1,2-dichlohloroethene; 6. 1,1,1-trichloroethane; 7. benzene; 8. trichloroethene; 9. toluene; 10. tetrachloroethene; 11. chlorobenzene; 12. ethylbenzene; 13. m-xylene & p-xylene; 14. o-xylene; 15. cumene (isopropylbenzene); 16. 2,4-dichlrobenzene-d_4 (surrogate); and 17. naphthalene at 2.6 ppm each was spiked in 0.7 mL water with 25% NaCl. The extraction was by headspace SPME for 12 sec using the 30 μm PDMS fiber. The analytes were desorbed for 2-3 min at 250°C into a low-volume injection port liner. The analysis was performed on an SPB-1 column 10 m x 0.20 mm x 1.2 μm that was heated from 70°C (0.2 min) to 180°C at 50°C/min. Hydrogen carrier gas at 12 psig was used with an FID detector.

Purge and trap, the traditional method to analyze VOCs, creates difficult conditions to achieve well-resolved, early-eluting peaks because the more volatile analytes are spread out while being transferred to the GC column. With SPME, no transfer line is used. Instead, the analytes must quickly desorb from the fiber and refocus at the head of the column. Using this technique, the key to obtaining sharp, well-resolved peaks is to reduce the volume of the inlet liner. This increases the effective linear velocity through the liner and reduces dead volume and band broadening; consequently, good chromatography of the initial eluting peaks is obtained [4]. Figure 3.8 compares results with a 0.75 mm ID liner to results with a standard 2 mm ID liner. The peaks are much sharper with the narrow bore liner. Even though these components are gases at room temperature, they are refocused at the head of the column without cryogenics as long as the narrow bore liner and a thick film capillary column are used.

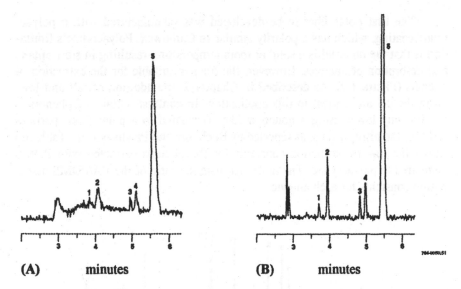

Figure 3.8 Comparison of inlet liners for the analysis of gaseous VOCs extracted by SPME. 1. chloromethane, 2. vinyl chloride, 3. bromomethane, 4. chloroethane and 5. Trichlorofluoromethane, at 50 ppb each in water, extracted by SPME and desorbed to **(A)** standard 2 mm ID splitless liner and **(B)** 0.75 mm ID SPME liner. The analytes were extracted (direct immersion) for 10 min using the 100 μm PDMS fiber. The analysis was identical for both runs using a 60 m x 0.25 mm x 1.2 μm VOCOL column. The oven was set at 40°C isothermal and the injection port temp was at 30°C with the splitter closed. Mass spectrometer was used as the detector with a scan range m/z = 45-280 at 0.6 sec/scan.

Thinner fiber coatings may also be advantageous if stirring or agitating is not possible because migration of semivolatiles into a thick film without agitation is slow and unpredictable. Even with hours of extraction, the uptake can be minimal and variable. A thinner fiber (7 μm) allows the analytes to diffuse into the layer more readily to quickly reach equilibrium.

In general, a thick fiber, such as a 100μm PDMS, is a good starting fiber for most applications. Nearly all analytes will be extracted to some extent. Detection levels and precision may require that a more specific fiber coating be used for optimization.

Effects of Fiber Polarity

The hypothesis "like prefers like" is the basis for the development of polar fiber coatings. Since polar analytes are water soluble, they are more difficult to extract. In fact, small polar analytes are usually the most difficult analytes to extract.

The first polar fiber to be developed was manufactured with a polyacrylate coating, which has a polarity similar to Carbowax. Polyacrylate's limitation is that the material is a solid at room temperature, resulting in slow uptake and desorption of analytes. However, the fiber is suitable for the extraction of phenols (Figure 3.9). As described in Chapter 2, the addition of salt and lowering the pH are critical to this application. In contrast, uptake of phenols is significantly lower using a nonpolar fiber compared to a polar fiber, particularly for the nitrophenols, as reported by Buchholz and Pawliszyn [5]. Table 3.6 shows the lower distribution constants for the phenols extracted with PDMS compared to polyacrylate. The table also lists the ratio of the PA/PDMS distribution constants for each analyte.

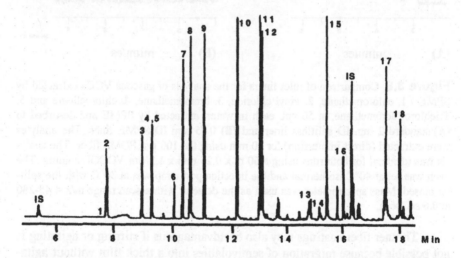

Figure 3.9 Analysis of phenols extracted from water by SPME. IS 2-fluorophenol; 1. phenol; 2. chlorophenol; 3. 2-methylphenol; 4. 3-methylphenol; 5. 4-methylphenol; 6. 2-nitrophenol; 7. 2,4-dimethylphenol; 8. 2,4-dichlorophenol; 9. 2,6-dichlorophenol; 10. 4-chloro-3-methylphenol; 11. 2,4,5-trichlorophenol; 12. 2,4,6-trichlorophenol; 13. 2,4-dinitrophenol; 14. 4-nitrophenol; 15. 2,3,4,6-tetrachlorophenol; 16. 2-methyl-4,6-dinitrophenol; IS 2,4,6-tribromophenol; 17. Pentachlorophenol; and 18. Dinoseb, at 50 ppb in water at pH = 2 and containing 25% NaCl. The extraction (direct immersion) is for 20 minutes using the PA fiber and analysis is on a PTE-5 column. The oven is programmed from 60°C (2 min) to 280°C at 10°C/min. The split/splitless injection port is set at 280°C and the fiber is desorbed for 5 min with the split vent closed. The detector is a quadrupole mass spectrometer with a scan range of m/z = 45-300 at 0.6 sec/scan.

Table 3.6 Distribution Constants for Phenols Using Two SPME Fiber Types.[a]

Component	Distribution Constants of SPME Fibers		
	PDMS	PA	PA / PDMS
4-nitrophenol	0.1	2.4	24.0
2-nitrophenol	0.2	3.7	18.5
2,4-dinitrophenol	0.1	1.7	17.0
2,4-dichlorophenol	4.6	47	10.2
2,4-dimethylphenol	1.3	9.1	7.0
4-chloro-3-methylphenol	2.4	16	6.7
2,4,6-trichlorophenol	15	60	4.0
2-chlorophenol	4.6	9.3	2.0
2-methyl-4,6-dinitrophenol	6	7.3	1.2
phenol	1.3	1.3	1
pentachlorophenol	370	170	0.5

[a] Lower values infer slower time to reach equilibrium.

The analyte order in the table is essentially the ranking of the analytes from high polarity to low. The exception is phenol, which is more polar than most of the substituted phenols. The high distribution constants indicate that the more polar fiber extracts the more polar analytes better than a nonpolar fiber. However, phenol was not well extracted by either fiber. Only the relatively nonpolar pentachlorophenol was extracted better by the nonpolar fiber. Linearity of the phenol extractions, as measured by percent relative standard deviation (% RSD), is shown in Table 3.7.

For most of the analytes, deviations are less than 15% from 5 ppb-200 ppb, with the exception of the very polar analytes. Because only a small portion of the nitrophenol and phenol molecules is extracted, quantitation at low ppb levels is difficult. Linearity improves dramatically for these analytes if the minimum quantitation level is 20 ppb.

Polar fibers were also shown to be advantageous in the work by Boyd-Boland and Pawliszyn on the extraction of nitrogen-containing pesticides [6]. The polyacrylate fiber was better for the extraction of these analytes than the 100µm PDMS fiber. In other work by Boyd-Boland involving 60 pesticides, the polyacrylate fiber again was the fiber of choice because it extracted the more polar triazines in lower concentrations than the PDMS fiber [7].

Table 3.7 Average Response Factors[a] and Linearity of Phenols.

Analyte	Average Response	% RSD
Phenol	0.33	11.4
2-Chlorophenol	1.15	4.6
2-Methylphenol	0.69	5.1
3- & 4-Methylphenol	1.18	6.9
2-Nitrophenol	0.29	25.6
2,4-Dimethylphenol	1.15	8.6
2,4-Dichlorophenol	1.32	10.8
2-4-Dichlorophenol	1.33	10.5
4-Chloro-3-methylphenol	0.98	7.9
2,4,5-Trichlorophenol	1.01	6.3
2,4,6-Trichlorophenol	1.01	8.4
2,4-Dinitrophenol	0.08	50.3
4-Nitrophenol	0.20	17.5
2,3,4,6-Tetrachlorophenol	0.70	18.7
2-Methyl-4,6-dinitrophenol	0.26	36.1
Pentachlorophenol	0.31	29.3
Dinoseb	0.27	29.1

[a] $Response\ Factor = \dfrac{\left(area\ counts_{analyte}\right) \times \left(concentration_{int.\ std.}\right)}{\left(area\ counts_{int.\ std.}\right) \times \left(concentration_{analyte}\right)}$

[b] Calculated over eight concentration points from 5 - 200 ppb. Internal standard is 2,4,6-tribromophenol at 50 ppb.

Either the polyacrylate fiber or the 100 μm PDMS fiber can extract the phosphorus containing pesticides (OPPs). Most of the OPPs are relatively non-polar; however, a few have significant polarity and are better extracted with the polyacrylate fiber. All of the pesticides extract rather slowly, especially with the polyacrylate fiber. A minimum extraction time is 30 minutes, but 45 minutes is more desirable. In all cases, 25-30% salt (NaCl) should be added to the solution. Direct immersion is required for most of these pesticides; however, heated headspace is suitable for some OPPs.

The effect of fiber polarity is demonstrated with the extraction of solvents, as seen in Table 3.8. These analytes at 500 ppb were extracted for 10 minutes, using heated headspace at 45°C. To enhance extraction efficiency, 25-30% NaCl was added. The two most polar analytes on this listing are methanol and ethanol. Relative to the nonpolar PDMS fiber, the polar polyacrylate fiber better extracts the two alcohols. The same correlation can be seen when comparing the relatively nonpolar PDMS-DVB fiber to the more polar CW-DVB fiber. The CW-DVB fiber better extracts alcohols, while the ketones are better extracted by the nonpolar fiber. However, all of these fibers extracted only a portion of the amount of analyte extracted by the Carboxen-PDMS fiber, indicating that the contribution from porosity is often greater than the contribution from polarity.

Table 3.8 Effects of Fiber Polarity on the Extraction of Solvents as Measured in Area Counts.

Component	PDMS 100µm	PA	CW DVB	PDMS DVB	Carboxen PDMS
Methanol	0	170	75	30	630
Ethanol	35	180	130	110	5300
Acetonitrile	140	230	130	160	6500
Acetone	400	260	250	640	97000
Isopropanol	180	360	250	600	57000
n-Propanol	220	1200	450	1200	83000
Ethyl acetate	1500	2700	4700	14000	450000
2-Methyl-3-propanone	4000	2100	13000	48000	820000

Effects of Fiber Porosity

The addition of a porous material suspended in a phase coating effectively increases the fiber film thickness. The improved extraction efficiency is due to the pores' abilities to retain analytes, even without increasing the film thickness. Figure 3.10 compares the uptake of C_2 - C_6 hydrocarbons by three fiber coatings. The 100 μm fiber does not retain the smaller hydrocarbons, but retains a small amount of pentene and hexene. The 65 μm PDMS-DVB fiber, which contains mesopores, better retains hexene and pentene than the 100 μm PDMS fiber; however, it does not extract ethylene or propene, and it extracts only a trace amount of butene. The highly microporous Carboxen-PDMS fiber retains the latter three analytes better and also extracts ethene, propene and butene. Table 3.9 lists the area counts for the alkane series, which responded similarly to the alkenes.

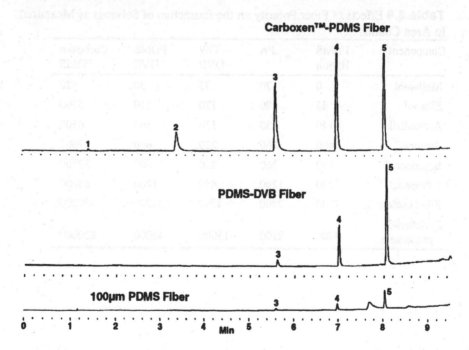

Figure 3.10 Comparison of hydrocarbon extractions using three different SPME fibers. The analytes are: 1. ethylene, 2. propene, 3. 1-butene, 4. 1-pentene, and 5. 1-hexene at 1 ppm in a gas bulb. The chromatograms are drawn at the same scale. The extraction is for 10 min in the gas phase with each fiber. The analysis is on a Supel Q™ column programmed from 40°C for 2 min to 200°C at 20°C/min. The injection port is set at 260°C for all of the fibers. The detection is by FID.

Table 3.9 Comparison of SPME Fibers for Extracting C_2 - C_6 Alkanes.[a]

Component	100µm PDMS	PDMS-DVB	Carboxen-PDMS
Ethane	0	0	750
Propane	0	0	20000
Butane	0	340	72000
Pentane	230	2150	108000
Hexane	460	9300	106000

[a] Response is similar to the alkene series described in Figure 3.10.

Uptake into the fiber increases as the hydrocarbons increase in size when using liquid phases. On the other hand, uptake of the C_4 - C_{12} analytes is identical in area counts with the Carboxen-PDMS fiber. This indicates that uptake of the analytes is not based on equilibrium, as with liquid phases. Once an analyte similar in size to C_4, or larger, is trapped in the Carboxen pore, it does not leave until the fiber is desorbed at a high temperature. Smaller analytes, with a diameter similar to ethane or propane, are subject to the equilibrium process (i.e., they can diffuse through, or out of, the pores). These analytes include monochlorinated and fluorinated C_1 hydrocarbons.

Prior to introducing the Carboxen-PDMS fiber, sulfur gases were extremely difficult to extract by SPME. Now, a variety of sulfur gases can be extracted, as shown in Figure 3.11. All of the gases were easily detected at 1 ppm v/v. Furthermore, hydrogen sulfide, sulfur dioxide, and the other sulfur gases can be detected at concentration levels of 10 ppb or less (w/v) or 50 ppb v/v using the specificity and sensitivity of a flame photometric detector (FPD).

The Carboxen-PDMS fiber is ideal for detecting SO_2 in food products such as wine, as shown in Figure 3.12. The concentration of sulfur dioxide in the wine was not labeled on the bottle, but it was estimated to be about 1%. This same fiber is very well suited for detecting the other volatile analytes in wine. By using the same extraction conditions, but changing to a mass spectrometer, volatile analytes in wine can be detected as shown in Figure 3.13. Sulfur dioxide was also detected by mass spectroscopy.

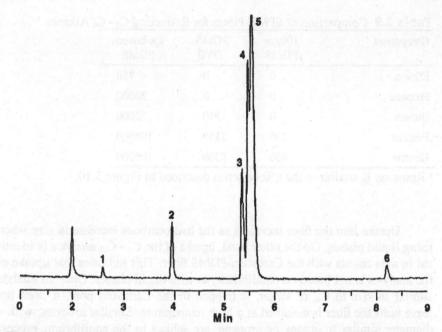

Figure 3.11 Analysis of sulfur gases extracted by SPME: 1. carbonyl sulfide, 2. methyl mercaptan, 3. ethyl mercaptan, 4. dimethyl sulfide, 5. carbon disulfide, and 6. dimethylsulfide are analyzed at 1 ppm v/v using a GC/MS quadrupole system. The extraction is with a Carboxen-PDMS fiber for 10 min in an air sampling bulb. The analytes are desorbed at 280°C for 2 min in a split/splitless liner that is closed for 2 min. The sample is delivered into a 30 m x 0.32 mm ID Supel Q™ column that is heated at 45°C (0.75 min) to 250°C at 25°C/min. Helium, with a linear rate set at 25 cm/sec, is used as the carrier gas.

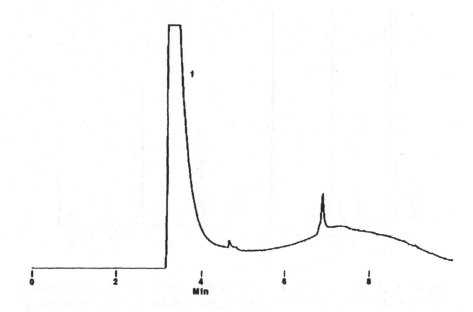

Figure 3.12 Analysis of white wine using SPME with FPD. One identifiable peak is detected with a FPD in the sulfur specific mode: sulfur dioxide. A Carboxen-PDMS fiber is used to extract wine containing 25% salt (NaCl). The extraction is done by heated headspace at 40°C for 10 min, and the fiber is desorbed at 280°C for 2 min into a split/splitless injection port containing a 0.75 mm ID liner. The split vent is closed for 1.5 min. Sulfur dioxide is not quantified. The Supel Q™ column used for the analysis is heated at 40°C for 1 min, then programmed to 270°C at 25°C/min.

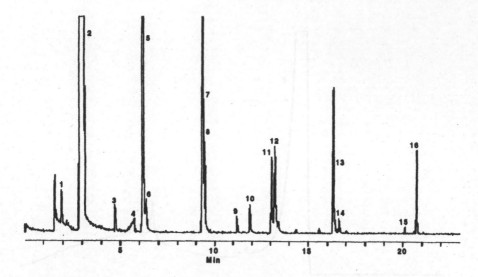

Figure 3.13 Analysis of white wine by GC/MS using SPME. The same extraction conditions used in Figure 3.12 were used for this analysis. The components detected are: 1. sulfur dioxide, 2. ethanol, 3. methyl formate, 4. acetic acid, 5. ethyl acetate, 6. isobutanol, 7. isopentanol, 8. 2-methyl-1-butanol, 9. ethyl butyrate, 10. 2,3-butanediol, 11. hexanol, 12. isoamyl acetate, 13. ethyl hexanoate, 14. hexyl acetate, 15. octanoic acid, and 16. ethyl octanoate, detected by quadrupole mass spectrometer with a scan range of m/z = 35-280 at 0.6 sec/scan. The analysis is carried out on a 30 m x 0.25 mm x 1.5 μm VOCOL column, programmed from 40°C (2 min) to 220°C at 10°C/min.

The effect of porosity is also demonstrated by the extraction of VOCs in Table 3.10, which compares Carboxen-PDMS and PDMS-DVB fibers to the 100μm PDMS fiber. The results are listed as ratios of responses for the listed fiber to the responses for the 100μm PDMS fiber. Thus, for the Carboxen-PDMS fiber, a value of 10 means that the Carboxen-PDMS fiber extracted 10 times more of that analyte than the PDMS fiber. The Carboxen-PDMS fiber strongly retained the early-eluting VOCs. Generally, the uptake using the Carboxen fiber was 10-100 times greater than for the 100μm PDMS fiber. The PDMS-DVB fiber was the best fiber for extracting the later eluting VOCs. However, the advantage of this fiber over the Carboxen-PDMS diminished when heated headspace was used. Here, a 2-10 fold increase in analyte response with heated headspace at 40°C was achieved, compared to immersion of the Carboxen-PDMS fiber.

Table 3.10 Relative Response of VOCs Using Carboxen PDMS and PDMS-DVB Fibers with Respect to 100 µm PDMS Fiber Response.

Component	Carbx-PDMS	PDMS-DVB	Component	Carbx-PDMS	PDMS-DVB
Dichlorofluoromethane	4	1	1,1,2-Trichloroethane	16	14
Chloromethane	10	1	Tetrachloroethylene	5	4
Vinyl chloride	47	3	Chlorodibromomethane	9	11
Bromomethane	314	196	Dibromoethane	13	13
Chloroethane	16	4	Chlorobenzene	4	4
Trifluoromethane	5	2	Ethylbenzene	3	4
1,1-Dichloroethene	18	4	o-Xylene	3	6
Methylene chloride	67	6	Styrene	4	5
1,1-Dichloroethane	35	12	Bromoform	4	12
2,2-Dichloropropane	3	4	Isopropylbenzene	2	4
cis-1,2-Dichloroethene	39	8	1,1,2,2-Tetrachloroethane	5	9
Trichloromethane	26	9	1,2,3-Trichloropropane	6	11
Bromochloromethane	63	122	Bromobenzene	4	4
1,1,1-Trichloroethane	4	4	1,3,5-Trimethylbenzene	2	4
1,1-Dichloropropene	10	4	2-Chlorotoluene	3	4
Carbon tetrachloride	4	4	tert-Butylbenzene	2	4
Benzene	17	6	sec-Butylbenzene	1	4
Trichloroethene	12	4	1,3-Dichlorobenzene	2	4
1,2 Dichloropropane	11	9	n-Butylbenzene	1	3
Bromodichloromethane	15	11	1,2-Dibromo-3-chloropropane	1	7
Dibromomethane	46	21	Hexachlorobutadiene	1	2
cis-1,3-Dichloropropene	11	8	Naphthalene	1	4
Toluene	7	5	1,2,3-Trichlorobenzene	1	4

The increased response enabled 60 VOCs in U.S. EPA Method 524.2 to be detected at 1 ppb (Figure 3.14); furthermore, most analytes were detected at the ppt level (Table 3.11). By using the Carboxen-PDMS fiber, SPME can meet minimum detection limits specified in the drinking water method. An actual drinking water sample is shown in Figure 3.15. Some of the disinfection by-products are detected in low ppt concentrations using ion trap mass spectrometry.

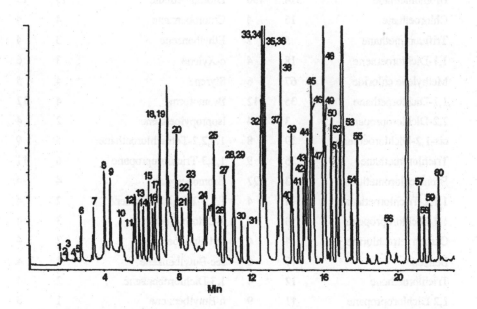

Figure 3.14 Analysis of U.S. EPA 524.2 VOCs from water extracted by SPME. The analytes listed in Table 3.11 are spiked at 1 ppb into water containing 25% NaCl, then extracted with a Carboxen-PDMS fiber by heated headspace at 40°C for 20 min. The fiber is desorbed for 5 min at 310°C in a split/splitless injection port containing a 0.75 mm ID liner. The analysis is on a 30 m x 0.25 mm x 1.5 µm VOCOL column heated at 40°C (2 min) to 210°C at 8°C/min. Detection is by ion trap mass spectrometry with a scan range of m/z = 45-260 at 0.6 sec/scan.

Table 3.11 Analytes in U.S. EPA Method 524.2.

Components		
1. Dichlorofluoromethane	21. 1,2 Dichloropropane	41. 1,1,2,2-Tetrachloroethane
2. Chloromethane	22. Bromodichloromethane	42. 1,2,3-Trichloropropane
3. Vinyl chloride	23. Dibromomethane	43. Propylbenzene
4. Bromomethane	24. cis-1,3-Dichloropropene	44. Bromobenzene
5. Chloroethane	25. Toluene	45. 1,3,5-Trimethylbenzene
6. Trifluoromethane	26. trans-1,3-Trichloropropane	46. 2-Chlorotoluene
7. 1,1-Dichloroethene	27. 1,1,2-Dichloroethane	47. 4-Chlorotoluene
8. Methylene chloride	28. Tetrachloroethylene	48. tert-Butylbenzene
9. trans-1,2-Dichloroethene	29. 1,3-Dichloropropane	49. 1,2,4-Trimethylbenzene
10. 1,1-Dichloroethane	30. Chlorodibromomethane	50. sec-Butylbenzene
11. 2,2-Dichloropropane	31. Dibromoethane	51. Isopropyltoluene
12. cis-1,2-Dichloroethene	32. Chlorobenzene	52. 1,3-Dichlorobenzene
13. Trichloromethane	33. 1,1,1,2-Tetrachloroethane	53. 1,4-Dichlorobenzene
14. Bromochloromethane	34. Ethylbenzene	54. n-Butylbenzene
15. 1,1,1-Trichloroethane	35. m-Xylene	55. 1,2-Dichlorobenzene
16. 1,1-Dichloropropene	36. p-Xylene	56. 1,2-dibromo-3-chloropropane
17. Carbon tetrachloride	37. o-Xylene	57. 1,2,4-Trichlorobenzene
18. Benzene	38. Styrene	58. Hexachlorobutadiene
19. 1,2-Dichloroethane	39. Bromoform	59. Naphthalene
20. Trichloroethene	40. Isopropylbenzene	60. 1,2,3-Trichlorobenzene

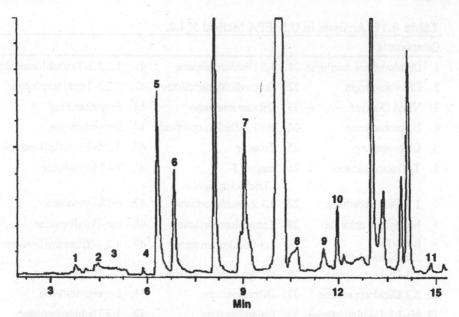

Figure 3.15 VOCs in drinking water sample by SPME-GCMS. Using the same conditions as in Figure 3.14, the following analytes are detected at listed concentration levels. 1. 1,1-dichloroethene (1 ppb), 2. methylene chloride (0.5 ppb), 3. trans-1,2-dichloroethene (0.03 ppb), 4. cis-1,2-dichloroethene (0.04 ppb), 5. Chloroform (15.2 ppb), 6. 1,1,1-trichloroethene (8 ppb), 7. bromodichloromethane (9.2 ppb) 8. toluene (2.1 ppb), 9. tetrachloroethene (0.4 ppb), 10. dibromochloromethane (8.1 ppb), and 11. bromoform (0.6 ppb ppb), detected at low ppt concentrations using ion trap mass spectrometry.

The presence of DVB also increases the polarity of the fiber, and surprisingly, DVB mixed with PDMS has a high affinity for amines. Extracting nitrogen containing analytes was enhanced with the PDMS-DVB fiber as shown in evaluating a small polar analyte mixture particular to the pharmaceutical manufacturing industry (U.S. EPA Method 1671) (Figure 3.16). The amount of amines extracted with respect to alcohols is significantly higher. Although the amines were present at 1/10 the level of the alcohols, response for the former was greater. SPME offers a great advantage over the current method that requires a direct aqueous injection, with minimum detection limits of 50 ppm or higher. For SPME extraction of the analytes, the pH of the water must be adjusted to 11 with a phosphate buffer and 25-30% NaCl is added. Both of these modifications are necessary to assure good reproducibility and linearity. The extraction can be accomplished by either direct immersion or heated headspace at 50°C. Both mass spectrometers and FIDs have been used as detectors for the analysis.

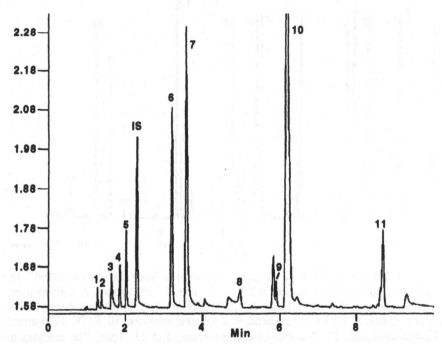

Figure 3.16 Analysis of U.S. EPA Method 1671 compounds extracted by SPME. The following analytes, at listed concentration levels, are extracted with the PDMS-DVB fiber from water adjusted to pH 11 = 11 with 0.05M phosphate buffer and 25% NaCl: 1. methylamine (1ppm), 2. methanol (10 ppm), 3. dimethylamine (1 ppm), 4. ethanol (10 ppm), 5. acetonitrile (10 ppm), IS. isopropanol (10 ppm), 6. n-propanol (10 ppm), 7. diethylamine (1 ppm), 8. 2-methoxyethanol (10 ppm), 9. ethylene glycol (25 ppm), 10. Triethylamine (1 ppm), and 11. dimethylsulfoxide (DMSO) (10 ppm). The sample is extracted by direct immersion for 12 min and desorbed for 4 min at 250°C. The analysis is performed on a 30 m x 0.32 mm x 4.0 μm SPB-1 Sulfur column programmed from 45°C for 2 min to 180°C at 10°C/min. The split/splitless injection port with a 0.75 mm ID liner is used with the split vent closed for the initial 2 min. The detector is an FID.

A Carbowax-DVB fiber is slightly better than a polyacrylate fiber for extracting solvents because of its more efficient release. Nevertheless, the Carboxen-PDMS fiber is best for extracting trace amounts (<100 ppb) of oxygenated solvents, as previously stated; although, this fiber will be overloaded in the ppm range. Hence, the Carbowax-DVB fiber is ideal for extracting these analytes in the mid ppb to upper ppm range.

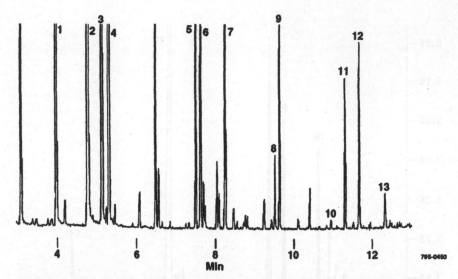

Figure 3.17 Analysis of nitroaromatic explosives in water extracted by SPME. The following explosives were present at 75 ppb in water: 1. nitrobenzene, 2. 2-nitrotoluene, 3. 3-nitrotoluene, 4. 4-nitrotoluene, 5. 1,3-dinitrotoluene, 6. 2,6-dinitrotoluene, 7. 2,4-dinitrotoluene, 8. 1,3,5-trinitrotoluene, 9. 2,4,6-trinitrotoluene, 10. RDX, 11. 4-amino-2,6-dinitrotoluene, 12. 2-amino-4,6-dinitrotoluene, and 13. Tetyl. The analytes are extracted by directly immersing the PDMS-DVB fiber into water adjusted to pH 9.5 containing 27% NaCl for 30 min. The analytes are desorbed at 260°C for 5 min into a split/splitless injection port. An amine-deactivated PTA-5™ 30 m x 0.25 mm x 0.25 μm column is programmed from 60°C (2 min) to 280°C at 10°C/min. Detection is by mass spectrometer with a scan range m/z = 45-350 at 0.6 sec/scan.

DVB also has a high affinity for aromatic compounds when combined with either Carbowax or PDMS. These fibers are suitable for extracting phenols, anilines, and a variety of pesticides. A particularly interesting analysis is the extraction of nitroaromatics (explosives) with the PDMS-DVB fiber. These analytes are normally analyzed by HPLC, but can be analyzed by GC, as shown in Figure 3.17.

SPME was coupled with a specific detector that does not utilize GC in work performed at the Jet Propulsion Laboratory in California. Minimum detection limits for many of the nitroaromatics were in the ppt range [8]. Desorption at 180°C greatly reduced thermal breakdown of some of the more labile analytes.

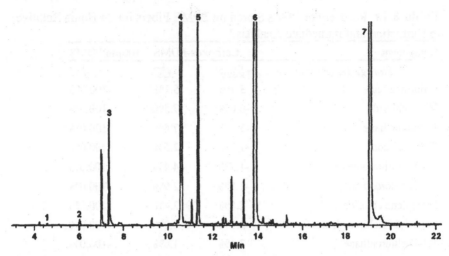

Figure 3.18 Analysis of nitrosamines at 10 ppb from water using SPME. The following analytes were at 10 ppb each in water adjusted to pH = 10 and containing 25% KCl: 1. nitrosodimethylamine, 2. nitrosodiethylamine, 3. nitrosomethylethylamine, 4. nitrosodipropylamine, 5. nitrosopiperdine, 6. nitrosodibutylamine, and 7. nitrosodiphenylamine. The analytes are extracted for 15 min by direct immersion with the PDMS-DVB fiber then desorbed at 270°C for 1 min. A PTA-5, 30 m x 0.32 mm x 0.5 μm column is used to separate the analytes. The oven is heated at 50°C for 1 min to 250°C at 10°C/min and held for 2 min. A mass spectrometer is used to detect the analytes in a selected ion mode.

Nitrosamines can be extracted at low ppb concentrations using the PDMS-DVB fiber. For example, Jeff Clark of the Liggett Group extracted a variety of nitrosamines from both water and tobacco. Figure 3.18 shows the analysis from water at 10 ppb for each analyte. The analytes could be detected to 1 ppb, but it was essential to have an amine-deactivated column to detect nitrosodimethylamine at this level. To extract the analytes, the solution was adjusted to pH 10 and was saturated with KCl.

SPME in Field Sampling and Air Monitoring

The ability to extract and maintain analytes on the fiber for an extended period of time enables SPME to be used in field sampling. Analytes can be extracted in the field and returned to the lab for analysis later, in order to reduce the chance of analytes breaking down in water. By sealing the end of the outer needle containing the fiber with a septum, the analytes can be stored on the fiber for extended periods of time.

Table 3.12 Recovery of VOCs Stored on SPME Fibers for 24 Hours Relative to Extraction and Immediate Analysis.[a]

Component	Carboxen-PDMS		100µm PDMS
Storage Temperature:	*Ambient*	*-4°C*	*-4°C*
Chloromethane	3.1%	5.1%	-100.0%
Vinyl chloride	-6.0%	7.5%	-100.0%
Bromomethane	-3.6%	7.4%	-100.0%
Chloroethane	-1.7%	2.5%	-100.0%
Trichlorofluoromethane	-1.7%	14.4%	-100.0%
1,1-Dichloroethene	-1.9%	2.9%	-100.0%
Methylene chloride	2.8%	2.4%	-100.0%
trans-1,2-Dichloroethene	-2.5%	2.2%	-100.0%
1,1-Dichloroethane	3.4%	1.6%	-100.0%
Chloroform	2.1%	-1.2%	-100.0%
1,1,1-Trichloroethane	-3.4%	4.5%	-100.0%
Carbon tetrachloride	6.1%	-1.6%	-100.0%
Benzene	-0.2%	0.9%	-98.3%
1,2-Dichloroethane	-14.8%	-14.0%	-94.2%

Table 3.12 shows U.S. EPA Method 624 analytes at 10 ppb that were extracted from water and successfully stored on the Carboxen-PDMS fiber for 24 hours as opposed to being immediately analyzed. One fiber was stored at ambient temperature, while the other fiber was stored at -4°C. The percent difference between the immediately analyzed sample and the stored sample is shown for each analyte. A negative value indicates a loss of analyte, and a positive value indicates a net gain due to variability among extractions. The results show that essentially no difference exists between samples stored on a Carboxen-PDMS fiber and samples immediately analyzed. Even the gaseous VOCs were not lost with storage in either cold or ambient conditions, which shows the strong capacity of the Carboxen-PDMS fibers to retain analytes. As previously described, the pore dynamics of Carboxen 1006 make the material a true adsorbent. In contrast, when these same analytes were extracted with a 100µm PDMS fiber and stored at -4°C, only the least volatile analytes were retained.

Table 3.12 Continued.

Component	Carboxen-PDMS		100µm PDMS
Storage Temperature:	*Ambient*	*-4°C*	*-4°C*
Trichloroethene	4.3%	12.2%	-100.0%
1,2-Dichloropropane	-8.1%	-1.1%	-97.5%
Bromodichloromethane	6.8%	-14.6%	-100.0%
Toluene	-8.2%	8.8%	-97.8%
trans-1,3-Dichloropropene	-6.2%	-3.5%	-98.3%
1,1,2-Trichloroethane	-5.0%	-6.9%	-100.0%
Tetrachloroethene	2.3%	-1.9%	-99.7%
Dibromochloromethane	4.2%	-2.9%	-99.5%
Chlorobenzene	1.2%	13.7%	-99.4%
Ethylbenzene	4.3%	12.1%	-99.5%
Bromoform	10.6%	-2.8%	-98.8%
1,1,2,2-Tetrachloroethane	-14.1%	-2.0%	-99.2%
1,2-Dichlorobenzene	11.5%	2.6%	-90.6%
Average	**-0.5%**	**1.9%**	**-98.1%**

$$^a\ \%\ Difference = \frac{\left(analyte\ response\ after\ storage\right) - \left(analyte\ response\ in\ immediate\ analysis\right)}{\left(analyte\ response\ in\ immediate\ analysis\right)}$$

The use of a Carboxen-PDMS fiber to monitor indoor air is shown in Figure 3.19. A 10-minute sampling of air in a research lab resulted in the detection of trace solvents (below 20 ppb) used for HPLC and GC applications. Martos and Pawliszyn have presented the ability to quantitate components in air using SPME [9]; furthermore, papers are being written that compare SPME with both active and passive air sampling monitors. Only SPME has the ability to sample accurately for 5-10 minute periods. In contrast, current NIOSH methods require monitoring for eight hours. Even for short sampling times, the minimum recommended time is 15 minutes, but one hour is preferred to obtain good sensitivity.

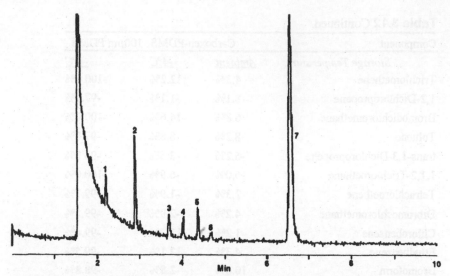

Figure 3.19 Solvents detected by monitoring the air in a research laboratory using SPME. A Carboxen-PDMS fiber was exposed for 10 min in the lab where the following solvents were detected: 1. methanol, 2. ethanol, 3. acetone, 4. acetonitrile, 5. methylene chloride, 6. hexane, and 7. isooctane. The fiber was desorbed at 300°C into a split/splitless injector with a 0.75 mm ID liner. The analysis was performed on a 30 m x 0.32 mm x 4.0 μm SPB-1 column programmed from 50°C (2 min) to 200°C at 10°C/min. Detection was by mass spectrometer with a scan range m/z = 45-260 at 0.6 sec/scan.

SPME field sampling of pesticides may be a better way to sample than shipping and storing samples in bottles for later analysis. Because the volatility of pesticides is significantly less than that of VOCs, a 100 μm PDMS fiber is suitable for storing extracted analytes. Table 3.13 outlines the results of a study comparing the stability of a variety of chlorinated, phosphorus-containing and nitrogen-containing pesticides stored on an SPME fiber, versus storage in containers, both over a 24 hour period at 4°C. Results were compared to values for a freshly prepared sample that was extracted by SPME and immediately analyzed.

Table 3.13 Recovery of Analytes Stored on an SPME Fiber or in Glass Vials.[a]

Analyte	Fiber[∇]	Water[♦]	Analyte	Fiber[∇]	Water[♦]
Atrazine	-15	-57	Methoxychlor	-14	-88
DDE	-12	-98	Methyl parathion	-7	-68
Disulfoton	-8	-93	Parathion	-15	-83
Endrin ketone	-10	-82	Phorate	-3	-84
Famphur	-3	-60	Simazine	-10	-53
Heptachlor epoxide	-12	-83	Sulfotep	+4	-81
Lindane	-2	-74	TEPP	-8	-54
Malathion	-6	-74	Thionazin	-3	-68
			Mean	**-8%**	**-75%**

[a] Measured as Percent Loss of Analyte

[∇] Pesticides at 10 ppb in water extracted by SPME and stored on fiber for 24 hours at 4°C.

[♦] Water sample stored in a silanized vial for 24 hours at 4°C, then extracted by SPME.

$$\% \text{ Difference} = \frac{\left(\text{analyte response after storage}\right) - \left(\text{analyte response in immediate analysis}\right)}{\left(\text{analyte response in immediate analysis}\right)}$$

A marked difference occurred between the types of storage when the results of the immediately analyzed sample were compared to the values for stored samples. An average loss of 8% of the analytes stored on the fiber occurred; whereas, a 75% loss occurred when stored in containers. Significant loss of analytes when stored in bottles was reported by Ferrer and Barcelo [10]. The loss of analytes may be due to breakdown or adsorption onto the glass surface. Effects of the type of container, pH, and temperature were also studied and were reported by Shirey [11]. The results indicate that extracting samples in the field may produce better results than collecting samples and returning them to the lab for later analysis.

A new SPME holder was designed for field sampling. Figure 3.20 shows this disposable unit with a built-in fiber. One or more of these units can easily be shipped in a cooler and stored in a refrigerator for later analysis. Although the data were obtained over a one-day period, studies up to two weeks have been made with no significant difference between 1 and 14 days.

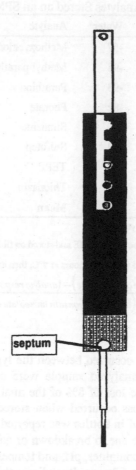

septum

Figure 3.20 Drawing of SPME portable field sampler.

Fiber Selection Guidelines

In many cases, several fiber options are available to extract a particular class of analytes; therefore, the concentration range becomes a critical factor. To illustrate, VOCs can be extracted by three different fibers: a Carboxen-PDMS fiber can extract ppt levels, but will be saturated in the ppm range; whereas, the PDMS 100 μm or 30 μm fibers can extract these analytes at the ppb or ppm range. Table 3.14 is a guide for choosing the proper fiber for a particular analyte class and concentration. Of course, the detector used also plays a role not only in the detection limits, but also in the linear range. For example, ECDs and ion trap detectors are very sensitive, which improves the minimum detection limits. On the other hand, they can be easily overloaded, which limits the upper end of the linear range. In this case, splitting the sample increases the linear range.

Table 3.14 SPME Fiber Selection Guide.

Analyte Class	Fiber Type	Linear Range
Acids (C2-C8)	Carboxen-PDMS	10 ppb-1 ppm
Acids (C2-C15)	CW-DVB	50 ppb-50 ppm
Alcohols (C1-C8)	Carboxen-PDMS	10 ppb-1 ppm
Alcohols (C1-C18)	CW-DVB	50 ppb-75 ppm
	Polyacrylate	100 ppb-100 ppm
Aldehydes (C2-C8)	Carboxen-PDMS	1 ppb-500 ppb
Aldehydes (C3-C14)	100 μm PDMS	50 ppb-50 ppm
Amines	PDMS-DVB	50 ppb-50 ppm
Amphetamines	100 μm PDMS	100 ppb-100 ppm
	PDMS-DVB	50 ppb-50 ppm
Aromatic amines	PDMS-DVB	5 ppb-1 ppm
Barbiturates	PDMS-DVB	500 ppb-100 ppm
Benzidines	CW-DVB	5 ppb-500 ppb
Benzodiazepines	PDMS-DVB	100 ppb-50 ppm
Esters (C3-C15)	100 μm PDMS	5 ppb-10 ppm
Esters (C6-C18)	30 μm PDMS	5 ppb-1 ppm
Esters (C12-C30)	7 μm PDMS	5 ppb-1 ppm
Ethers (C4-C12)	Carboxen-PDMS	1 ppb-500 ppb
Explosives (Nitroaromatics)	PDMS-DVB	1 ppb-1 ppm
Hydrocarbons (C2-C10)	Carboxen-PDMS	10 ppb –10 ppm
Hydrocarbons (C5-C20)	100 μm PDMS	500 ppt-1 ppm
Hydrocarbons (C10-C30)	30 μm PDMS	100 ppt-500 ppb
Hydrocarbons (C20-C40+)	7 μm PDMS	5 ppb-500 ppb
Ketones (C3-C9)	Carboxen-PDMS	5 ppb-1 ppm
Ketones (C5-C12)	100 μm PDMS	5 ppb –10 ppm
Nitrosamines	PDMS-DVB	1 ppb-200 ppb
PAHs	100 μm PDMS	500 ppt-1 ppm
	30 μm PDMS	100 ppt-500 ppb
	7 μm PDMS	500 ppt-500 ppb

Table 3.14 Continued.

Analyte Class	Fiber Type	Linear Range
PCBs	30 µm PDMS	50 ppt-500 ppb
Pesticides, chlorinated	100 µm PDMS	50 ppt –500 ppb
	30 µm PDMS	25 ppb-500 ppb
Pesticides, nitrogen	Polyacrylate	50 ppt-500 ppb
Pesticides, phosphorus	100 µm PDMS	100 ppt –1 ppm
	Polyacrylate	100 ppt-500 ppb
Phenols	Polyacrylate	5 ppb-500 ppb
Surfactants	CW-TPR	1 ppm-100 ppm
Sulfur gases	Carboxen-PDMS	10 ppb-10 ppm
Terpenes	100 µm PDMS	1 ppb-10 ppm
VOCs	Carboxen-PDMS	100 ppt-500 ppb
	100 µm PDMS	20 ppb-50 ppm
	30 µm PDMS	100 ppb-50 ppm

SPME AND HPLC

Hardware

In the past, the initial capabilities of SPME were limited in the ability to extract nonvolatile and thermally labile analytes. Two approaches exist for sampling these types of samples with SPME: (1) sample derivatization to make it more volatile and GC compatible, or (2) method development for solvent desorption and coupling SPME with HPLC analysis. Okyeo and Snow at Seton Hall University [12] and Pan and Pawilszyn at the University of Waterloo [13] have documented derivatization using SPME. Although derivatization may produce good results and does not greatly complicate the analysis, many analysts would prefer not to derivatize and use HPLC instead. Secondly, using solvents to desorb analytes from an SPME fiber is in its initial stages. SPME can be coupled with HPLC if analytes can be desorbed in a small solvent volume. The initial work by Chen and Pawliszyn coupled a low volume tee to an HPLC valve and demonstrated the feasibility with the analysis of PAHs [14].

Supelco introduced a commercial HPLC interface that contains a low volume cell (70 µl desorption chamber) in 1996. Sealing is accomplished by passing the fiber through a ferrule and clamping the ferrule against the inner tubing of the SPME fiber assembly (Figure 3.21).

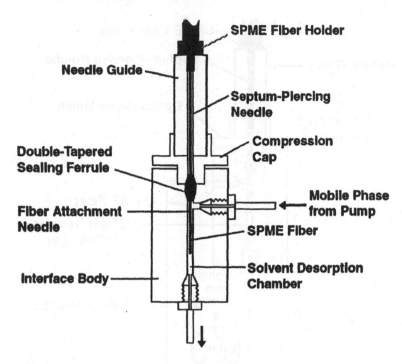

Figure 3.21 Schematic view of the SPME-HPLC interface.

The desorption chamber is attached to a 6-port valve that can be switched to allow the fiber to soak in solvent, then it is switched back to pass the mobile phase through the chamber. This sweeps the analytes into the column for analysis (Figure 3.22). In nearly all of the applications developed to date, allowing the fiber to soak for several minutes (static mode) is recommended over dynamically desorbing the analytes from the fiber. Additionally, the valve contains a port that allows you to fill the chamber with an alternative, stronger solvent that may better desorb the analytes from the fiber than the mobile phase. In typical reversed phase chromatography, the desorption solvent is stronger (less polar) than the initial gradient concentration of the mobile phase. Sometimes this can broaden peaks, but when standard size HPLC columns are used, the analytes will usually focus on the head of the column.

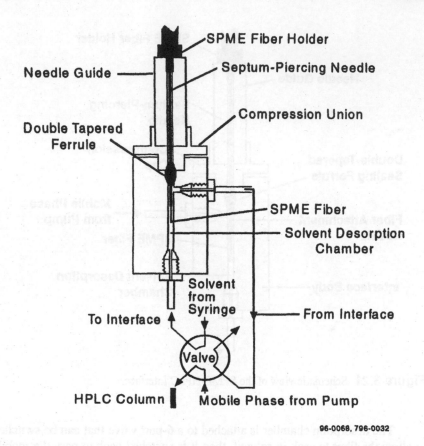

SPME Fiber Holder

Needle Guide

Septum-Piercing Needle

Compression Union

Double Tapered
Ferrule

SPME Fiber

Solvent Desorption
Chamber

Solvent
from
Syringe

To Interface

From Interface

Valve

HPLC Column

Mobile Phase from Pump

96-0068, 796-0032

Figure 3.22 Schematic view of HPLC interface and valve set for static desorption.

Because the analytes can be desorbed in a small volume of solvent, desorption off-line is possible, then the entire volume is injected through an HPLC loop. A manifold is being designed that will allow multiple extractions to be made simultaneously. These desorptions could be made in small vials, then the vials could be placed in an HPLC autosampler and injected. This would greatly reduce time relative to the single extraction and desorption method using the HPLC interface.

SPME Fibers Used in HPLC Applications

The current fiber technology for HPLC applications consists primarily of modifications of fibers for GC use. However, new problems were encountered with HPLC that were not encountered in GC use. For example, more fiber damage occurred, mostly due to fiber phase swelling. To illustrate, when the fiber was retracted through the ferrule after desorption, the phase would be stripped off the fiber if swelling occurred. The fiber could also break if it was not properly inserted through the ferrule. Another problem was due to the glue used to hold the fiber in the inner needle. Even though the glue was solvent resistant, some extraneous peaks were detected when certain solvents were used.

To help reduce these problems, the fused silica fibers were drawn with a plastic-like pre-coating, then coated with the standard phases. The plastic coating offered several advantages: (1) The fibers were extremely flexible and very unlikely to break. (2) The phase bonded better to the plastic than to bare fused silica, which greatly reduced swelling and stripping. (3) The fiber could be crimped into the inner tubing, thus eliminating the need for glue. Crimped fibers are not suitable for GC use because water vapor tends to wick into the inner tubing if the fiber is not glued.

At this time, two HPLC-specific fibers are commercially available: PDMS-DVB and CW-TPR, both on the plastic coating. Additionally, two other fibers are commonly used for HPLC. These are the 85 μm polyacrylate fiber and the 100 μm PDMS fiber.

Applications using HPLC started with analyses that were previously proven by GC. These include PAHs [14] and nitroaromatic explosives [15]. The areas of interest, however, are analytes that cannot be analyzed by GC. For example, work by Torbin Nilsson of the European Joint Research Commission details the analysis of acid herbicides using the PDMS-DVB-HPLC fiber [16].

Additionally, K. Jinno and associates of Toyohashi University of Technology have developed and published most of the clinical applications for SPME-HPLC to date. The researchers' primary work was with benzodiazepines from biological fluids [17]. Other applications published include the analysis of pesticides in biological fluids [18]. In all cases, Jinno used the 85 μm polyacrylate fiber. Extraction times were typically 1 hour or more and desorption times ranged from 30 minutes to 1 hour. Jinno is now investigating the use of a thinner polyacrylate phase to shorten sample preparation time.

A. Boyd-Boland investigated the use of SPME for extracting surfactants from water and determined that the CW-TPR was the best fiber for this analysis [19]. G. Gora-Maslak of Supelco has developed applications for tricyclic antidepressants (Figure 3.23).

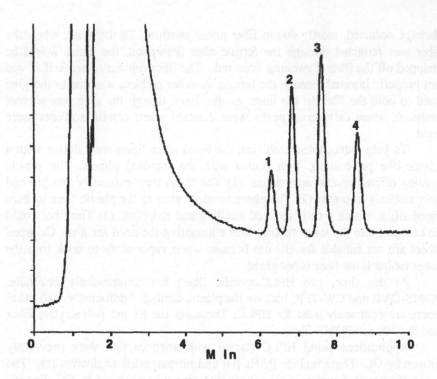

Figure 3.23 Analysis of tricyclic antidepressants from serum by SPME-HPLC. The following components are spiked into serum at 100 ng/ml: 1. nordoxepin, 2. nortripyline, 3. desipramine, and 4. protriptyline (IS). Serum is spiked with 20% methanol by volume to denature protein. The analytes are extracted with the PDMS-DVB-HPLC fiber for 30 min by directly immersing the fiber, then statically desorbed for 10 min in methanol:acetonitrile (60:40) and swept into a 25 cm x 4.6 mm ID SUELCOSIL™ LC-PCN column with acetonitrile:water:0.01M phosphate buffer pH = 7 (60:15:25). Detection is by UV at 215 nm. The precision is good at less than 10% RSD, and the linear range from 50-400 ng/ml has a correlation value of 0.999+.

Other applications include analyzing urea-containing pesticides, which were detected down to 1.5 ppb (Figure 3.24) and Triazine herbicides at 10 ppb [20]. In each case, the fiber of choice was the PDMS-DVB-HPLC fiber.

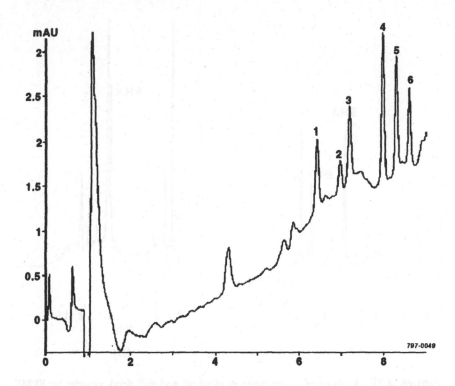

Figure 3.24 Analysis of urea-containing carbamates at 1.5 PPB by SPME-HPLC. The analytes extracted are: 1. Propham, 2. Siduron, 3. Linuron, 4. Chlorpropham, 5. Barban, and 6. Neburon, all at 1.5 ppb in water with 10% NaCl. The analytes are extracted with the PDMS-DVB-HPLC fiber for 40 min by direct immersion, then statically desorbed in acetonitrile:water (65:35) for 5 min and dynamically swept onto a 15 cm x 4.6 mm ID SUPELCOSIL LC-8 column with a mobile phase gradient of acetonitrile:water starting at 18:82 to 65:35.

SPME-HPLC has also been used for food analyses. These applications include antioxidants in oils and benzoic acid in colas. In both cases, the PDMS-DVB HPLC fiber was determined to be most suitable. Figure 3.25 shows the analysis of the antioxidants in oil and soft drink powder.

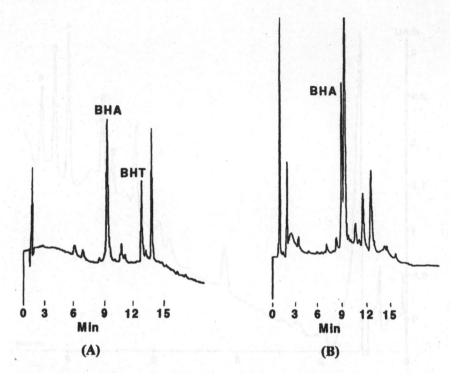

Figure 3.25 Analysis of antioxidants in olive oil and soft drink powder by SPME-HPLC. The analytes are extracted from oil **(A)** at 1 mg/g of each analyte by immersing the PDMS-DVB-HPLC fiber in the oil and extracting for 15 min. Antioxidants at 1 mg/g contained in 0.6 g of soft drink powder **(B)** dissolved in 0.3 ml of water are extracted for 20 min with the same fiber. The analytes are removed from the fiber by static desorption for 5 min in acetonitrile: methanol (50:50), then swept onto the column with a mobile phase of (A) 5% acetic acid in water with acetonitrile:methanol (1:1) (B) in a gradient starting with 30% and increasing to 100% over a 10 min period.

EXTENDING FIBER LIFE

Usually, a fiber will break, the coating will be stripped off, or the needle will be severely bent before fiber life is expended from normal use. Typically, the needle is bent when it is piercing a vial septum. The Teflon coating on the septum can vary in thickness depending upon the septum type. To reduce needle bending on a manual holder, adjust the depth gauge (needle guide) on the SPME holder so that only 0.5-1 cm of the needle is exposed. Place the holder on top of the vial septum and carefully turn the needle gauge to extend the fiber through the septum. Once it is through the septum, adjust the needle to proper depth. A popping sound will be heard when the needle penetrates the septum. This procedure is not needed with the automated SPME holder.

Phase stripping from the fiber is probably the second highest cause of fiber damage. This happens because the fiber coating swells or is damaged by the injection port septum. Fiber swelling does not occur unless the fiber is exposed to high concentrations of a nonpolar solvent. For example, this can occur in heated headspace extraction of a water sample containing high ppm or percent levels of gasoline, hydrocarbons, or ethers. The vapor is primarily organic solvent that can swell the fiber. When the fiber is retracted, the outer sheath will strip off the coating. This problem generally occurs with the 100 μm PDMS fiber and sometimes with the CW-DVB fiber. In HPLC use, introducing water into the desorption chamber after the sample has been delivered into the column can reduce fiber swelling and stripping. Water helps to shrink the phase coating and reduce the chance of stripping the fiber coating when it is retracted through the ferrule. In the SPME-HPLC interface, the valving can be switched for the mobile phase to by-pass the interface. At this time, the water can be entered through the waste port.

The inlet septum is always a problem, particularly if the septum nut is too tight. In this case, the needle is pushed through an over-stressed septum forcing the septum into the needle opening. If a piece of septum is lodged in the needle, the fiber will be forced into the septum. This will either strip the coating off the fiber or bend the inner tubing plunger. If a fiber becomes stripped, do not try another fiber before checking the liner for septum particles and loosening the septum nut. Furthermore, septum particles in the injector liner can damage the fiber and also cause baseline bleed problems. Therefore, liners should be checked for particles before the desorption step. Septum coring can be reduced by using pre-drilled septum and by not over-tightening the septum nut. Ultimately, the best solution is to eliminate the septum and use a septumless injection port, such as a Merlin Microseal. (Refer to the section on future fiber trends and Chapter 2).

Additional problems can occur from salt buildup in the injection port liner if the liquid sample remains on the SPME outer needle after it is removed from the sample container. This is potentially a problem when manually sampling from an open container, but it should not occur when sampling from a closed vial because the septum will wick off the liquid as the sheath is removed from the vial. When removing a fiber from a solution containing high concentrations of salt, confirm that water droplets are not visible on the outer needle. If they are, wipe the water off with a tissue before inserting it into the injector.

A dirty fiber will have reduced capacity for adsorbing analytes; consequently, a loss in sensitivity may occur during a sequence of analyses. Usually, thermal desorption will sufficiently clean the fiber, although prolonged thermal cleanings may be required. If a fiber is extremely dirty, it can be rinsed in a polar organic solvent such as methanol. However, this approach should be considered a final resort.

FUTURE FIBER TRENDS

The future of SPME fiber technology is difficult to speculate. Certainly, more fibers for both GC and HPLC applications are needed. Moreover, to make SPME compatible with new technologies, changes in the fiber assembly design may need to occur. For example, a new line of SPME fibers that are compatible with the Merlin Microseal is being developed. This seal requires a 23-gauge needle with a polished tip in contrast to the standard SPME outer needle that is 24-gauge. Eliminating the injection port septum reduces fiber damage and eliminates extraneous peaks caused by septum particles in the liner. The 23-gauge needles will be identified by a green sealing septum instead of the standard gray septum.

Another design change that has been available on request is a 2 cm fiber (instead of the standard 1 cm) to increase the fiber coating volume and, theoretically, extract twice as much sample. This longer fiber is assembled by cutting off 1 cm of the inner tubing to enable a 2 cm fiber to be inserted. To retract and fully expose the fiber, a springless fiber assembly (automated version) must be used.

Several fibers are being investigated as potential applications for GC. First, a Nafion-coated fiber has been investigated for the extraction of polar analytes from a nonpolar matrix. This extremely polar fiber tends to exclude nonpolar analytes, which makes it suitable for extracting oxygenates from fuels. Secondly, cyclodextrins are also being investigated for trapping polar analytes that are difficult to extract on existing fibers. In addition, other Carboxen fibers with different pore sizes are being researched for extracting fixed gases. Other small analytes of interest are ethylene and methane.

The foremost area of interest for HPLC is ion exchange fibers. A fiber that would allow desorption by increasing the salt concentration is desirable because it would reduce fiber swelling and may increase the rate of desorption. Another interesting potential fiber coating is a sol/gel (porous silica). This material has a polarity similar to Tenax® TA, but is slightly more microporous. The fiber appears to have properties similar to 100 μm PDMS, but is slightly more polar. The polarity of the sol/gel can be altered by derivatizing with longer carbon chain groups that may be compatible with HPLC applications. Finally, the possibility of incorporating antibodies onto the fiber would make it feasible to have specific-molecule fiber applications. However, this may be extremely difficult due to the limits of fiber capacity.

REFERENCES

1. KSW Sing, DH Everett, RAW Haul, L Moscou, RA Pierotti, J Rouquerol, and T Siemieniewska. Reporting physisorption data for gas/solid systems with special reference to the determination of surface area and porosity. Pure & Appl Chem 57:603-619, 1985.

2. RE Shirey, V Mani, and WR Betz. New carbon-coated Solid Phase Microextraction (SPME) fibers for improved analyte recovery. Presentation Pittsburgh Conf, 1997, Supelco, Publication T495015:1-17, 1997.

3. T Schumacher. Fast screening of water and soil samples using Solid Phase Microextraction (SPME) Supelco Reporter 16, (1):8, 1997.

4. RE Shirey. Rapid analysis of environmental samples using Solid Phase Microextraction (SPME) and narrow bore capillary columns. J High Res Chromatogr 18:495-499, 1995.

5. K Buchholz and J Pawliszyn. Optimization of solid-phase microextraction conditions for determination of phenols. Anal Chem 66: 160-167, 1994.

6. A Boyd-Boland and J Pawliszyn. Solid-phase microextraction of nitrogen containing herbicides. J Chromatogr 704: 163-169, 1995.

7. A Boyd-Boland, S Magdic, and J Pawliszyn. Simultaneous determination of 60 pesticides in water using solid-phase microextraction and gas chromatography. Analyst 121:929-938, 1996.

8. M Durrach personal communication, 1996.

9. P Martos and J Pawliszyn. Calibration of Solid Phase Microextraction for air analyses based on physical chemical properties of the coating. Anal Chem 69:206-215, 1997

10. I Ferrer and D Barcelo. Stability of pesticides stored on polymeric solid-phase extraction cartridges. J Chromatogr 778:161-170, 1997.

11. RE Shirey. Field sampling of pesticides using Solid Phase Microextraction. Supelco Publication T497170:1-15, 1997.

12. P Okeyo, S Rentz, and N Snow. Analysis of steroids from human serum by SPME with headspace derivatization and GC. J High Res Chromatogr 20:171-173, 1997.

13. L Pan, M Chong, and J Pawliszyn. Determination of amines in air and water using derivatization combined with SPME. J Chromatogr 773:249-260, 1997.

14. J Chen and J Pawliszyn. Solid-phase microextraction coupled to high performance liquid chromatography. Anal Chem 67:2530-2533, 1995.

15. RE Shirey, G Gora-Maslak, and V Mani. New Solid Phase Microextraction fibers and applications for GC and HPLC. Eastern Anal Symp Presentation. Supelco Publication T496170:1-18, 1996.

16. T Nilsson, D Baglio, I Galdo-Miguez. Solid-phase microextraction for the analysis of phenoxy acid herbicides. 20th International Symposium on Capillary Chromatography, Riva del Garda, Italy, 1998.

17. K Jino, M Taniguchi, H Sawada, M Hayashida. Microcolumn liquid chromatography coupled with solid-phase microextraction (SPME/Micro-LC) for the analysis of benzodiazepines in human urine. Analusis 26(5): M27-M30, 1998.

18. K Jino, T Muramatsu, T Saito, Y Kiso, S Magdic, and J Pawliszyn. Analysis of pesticides in environmental water samples by SPME-HPLC. J Chromatogr 754:137-144, 1996.
19. A Boyd-Boland and J Pawliszyn. SPME coupled with HPLC for determination of alkylphenol ethoxylate surfactants in water. Anal Chem 68:1521-1525, 1996.
20. G Gora-Maslak and V Mani. New generation of fibers for SPME/HPLC applications. Presentation Pittsburgh Conference, 1997. Supelco Publication T497027: 1-12, 1997.

4
Pharmaceutical Applications

Stephen Scypinski
The R.W. Johnson Pharmaceutical Research Institute, Raritan, New Jersey

Ann-Marie Smith
Hoffmann-LaRoche, Inc., Nutley, New Jersey

INTRODUCTION

The world of pharmaceutical analysis is very heavily regulated and is indeed "where good science and compliance coexist." Many conditions influence how pharmaceutical analytical chemists conduct their daily activities, whether they are developing the product itself or using methodology for the testing and release of a marketed product. These conditions are mandated by the Food and Drug Administration (FDA) [1], various international regulatory authorities, and the International Conference on Harmonization (ICH) [2]. Adamovics has edited a text discussing the impact of regulatory scrutiny on pharmaceutical analysis from the chromatography perspective [3]. A detailed discussion concerning development and validation of analytical methods is also available [4].

Solid Phase Microextraction holds great promise for the pharmaceutical analysis laboratory. Indeed, SPME has been heralded as one of the most innovative new techniques in the area of solvent-free sample preparation [5]. Because it is such a new technique, a great deal of prior art is not currently available. Therefore, in considering the challenge of preparing this chapter, we have presented the most obvious application of SPME first, which is determining solvent residues in drug substances. Lastly, we have attempted to summarize other work in the pharmaceutical analysis arena.

IMPURITIES AND RESIDUAL SOLVENTS

Analytical chemistry applied to the pharmaceutical industry is the process of determining the quality, purity, and potency of a particular test article, be it a raw material, active ingredient, excipient, final dosage form, or package component. As part of this process, determining impurities is one of the most crucial tests performed during development, as well as after the product has been approved for marketing clearance by the various regulatory bodies. In this regard, one facet of regulatory approval for a pharmaceutical dosage form is the inclusion of one or more validated analytical methods to control both the presence of identified (or specified) impurities, as well as tests and procedures that have a high probability of detecting unidentified impurities. Many opportunities occur for impurities to make their way into the pharmaceutical product because it typically undergoes a long and involved series of manufacturing steps, starting with the synthesis of the active ingredient and culminating with the final packaging steps for the dosage form.

One step identified as critical to the purity (and possibly efficacy and safety) of the final product is the synthesis of the active ingredient. Depending on the complexity of the synthesis, a considerable number of solvents and other volatile materials, which are conventionally referred to as Organic Volatile Impurities (OVIs), may be employed. Even after recrystallization and drying, many products contain trace levels of residual solvents in the ppb or ppm range. The amount of the final recrystallization solvent may be higher, often in the tenths of a percent range. As a result, products must be tested for OVIs and recrystallization and other toxic solvents.

Organic Volatile Impurities are specifically five toxic solvents that are monitored in drug substances and raw materials. The solvents are methylene chloride, chloroform, benzene, trichloroethylene, and 1,4-dioxane. The United States Pharmacopeia (USP) National Formulary (NF) documents the test procedures needed for OVI analysis in USP test <467>[6]. Although residual solvents, such as those involved in recrystallization and purification procedures, have been routinely monitored by such techniques as gas chromatography (GC) for a considerable time, the introduction of USP <467> has, over the last ten years, caused a volume of crosstalk about the proper control of volatile solvents. Adhering to the method specified in the pharmacopeial monograph can, and often will, lead to falsely identifying a common processing solvent as an OVI because the individual monographs listed in the USP do not reference specific manufacturers or the process that may be used in the synthetic procedure [7]. The ICH has also become involved and a recent review gives the status of the various guidelines proposed for the control of OVIs [8].

ANALYSIS OF ORGANIC VOLATILE IMPURITIES AND OTHER SOLVENTS USING SPME

Instrument Considerations for Determining Residual Solvents

Because residual solvents are controlled during the testing of active pharmaceutical substances, raw materials, and excipients, the remainder of this discussion will focus on these specific items. Several commonly employed techniques have been used for separating, measuring, and, in some cases, concentrating residual solvents in pharmaceutical ingredients. These include direct injection in aqueous and organic matrices, dynamic and static headspace, purge-and-trap, and, most recently, SPME. As previously mentioned, individual requirements that depend quite heavily on the synthetic procedure govern the methodology that will be utilized for determining solvent levels in active substances. To this end, no single method is suitable for all purposes due to the complexity of the sample matrix. The proper choice of solvent method must take into account the volatility of the solvents in question, solubility of the test article in water or an alternate solvent system, thermal lability of the substance, polarity of the solvents, and the detection limits desired. For some active substances and raw materials, direct injection will suffice [9-11]; however, a great majority of pharmaceutical analysis laboratories are currently moving toward either static or dynamic headspace sampling as the injection mode of choice [12-14].

SPME offers a new option for the rapid sampling and possible concentration of solvents from active substance matrices. This technique, originally developed for environmental analysis by Pawliszyn and co-workers [15], has found a variety of uses in many and varied application areas [16-17]. A recently published text on the subject details its many uses [18]. Despite the countless references to the environmental, chemical, and forensic areas, little work has been done in applying SPME to pharmaceutical analysis. Indeed, the reference section of the text discussed above groups citations to pharmaceutical SPME applications with those involving flavors, foods, and natural products. Nonetheless, SPME was compared and contrasted to the two commonly accepted sampling modes for analyzing residual solvents with respect to critical validation parameters and analytical figures of merit [19].

As the majority of the industry is gravitating toward the use of wide-bore or "Megabore" fused silica capillary columns for GC analysis, the work discussed here employed such columns exclusively. These columns can be used in conjunction with direct, on-column, and split/splitless capillary injection techniques for sample introduction. On-column injectors place the sample directly in the column, thereby providing the advantages of lack of mass discrimination, minimization of decomposition of thermally labile compounds, better precision, and hence, improved quantitation. However, the traditional on-column inlet does not enhance vaporization. Moreover, on-column analyses may be confounded by the presence of the active compound(s), or a sample matrix that may produce an abundance of interfering peaks, or column bleed upon decomposition of the non-volatile materials when the temperature program is initiated [20]. Nevertheless, direct injection has been the method of choice for most pharmaceutical applications because this inlet system is most amenable to the analysis of polar and thermally labile materials [21]. Although one is not typically interested in assaying for the active substance when conducting solvent analyses, the use of direct injection allows minimum decomposition of the active, thereby resulting in a cleaner chromatogram. However, special considerations in the use of direct injection must be understood. These include the use of high capacity GC columns, silanization, and packing of the inlet liner or sleeve [7, 20].

Split/splitless injectors, in their various designs, can be used to concentrate trace components onto a fused silica capillary column. Both injection modes require a certain skill level and considerable method development time to optimize the conditions for reproducible results. Split/splitless injectors are not routinely utilized in pharmaceutical analysis, but rather are reserved for such complex analyses as those performed in petrochemical and environmental laboratories [20,21].

Static headspace sampling has been used with great success in pharmaceutical solvent analysis as an alternative to the aforementioned injection modes. Headspace provides a sample introduction mechanism that uniformly heats and transfers the volatile portions of a sample directly into a GC column [22]. The obvious advantages are the virtually complete elimination of any nonvolatile materials, leading to better precision, low mass discrimination, cleaner chromatograms, and longer column life. Additionally, static headspace offers the possibility of sampling from aqueous systems, which cannot be typically done with direct, on-column, or split/splitless systems. Nevertheless, its principal disadvantage, especially to a routine analysis laboratory, is the necessity for specialized instrumentation.

Based on this knowledge and prior art in the field of GC analysis for solvent residues and OVIs, one can project that SPME could offer potential benefit for this application since the use of standard instrumentation would be allowable. With the proper method development and choice of fiber, it should be possible to use this new sampling tool to effectively discriminate between the solvents and the active substance.

Residual Solvent Analysis:
SPME vs. Direct Injection and Headspace Sampling

Early work from the laboratory of Clark, Scypinski, and Smith examined the SPME technique for a variety of processing solvents and OVIs [19], comparing it to both direct injection and headspace sampling with regard to validation criteria and analytical figures of merit. Although SPME is also available as a manual syringe type assembly from Supelco [23], the work described here was performed with the automated version of SPME first described several years ago by Arthur and co-workers [24] and marketed by Varian Chromatography Systems. The reason for this was to critically evaluate whether SPME holds merit for a pharmaceutical analysis laboratory performing routine testing because virtually all "Quality Control" instrumentation in the pharmaceutical industry is automated.

The solvents investigated were methanol, ethanol, acetone, isopropanol, methylene chloride, chloroform, benzene, trichloroethylene, and 1,4-dioxane. Some of the variables considered in this study were: (1) the use of SPME sampling in the headspace of the sample vial versus liquid sampling, (2) the use of salt as an equilibrium shifting agent to increase the partitioning of the solvents onto the SPME fiber, (3) SPME sampling from a methanolic solvent system versus an aqueous solution, and (4) the chemical nature of the fiber. Our research indicated that the latter variable played a significant role, as would be expected. This is shown in Figure 4.1, which depicts gas chromatograms of the OVIs and several common processing solvents on three of the more commonly used fibers. As illustrated, the selectivity of the various fibers varies with the polarity of the solvent in question. The more nonpolar (poly)dimethylsiloxane exhibited reasonable recoveries for all solvents in this study, with the exception of methanol. As methanol is an important solvent for pharmaceutical processing, alternate fibers were investigated for their suitability to adsorb the solvents of interest. The more polar polyacrylate fiber showed increased recovery for methanol, ethanol and, to a lesser extent, chloroform; however, relatively poor recovery for 1,4-dioxane was obtained using this fiber phase. Carboxen proved to be an even poorer choice of SPME fiber as nearly no recovery was observed for methanol, chloroform, and 1,4-dioxane. Therefore, all future experiments were performed using the (poly)dimethylsiloxane fiber.

Table 4.1 tabulates the system precision data obtained using the PDMS fiber and the GC conditions given in Figure 4.1. All three injection modes are suitable because the USP requirement for system precision of OVIs is % RSD values not greater than 15%. Note that SPME compares quite favorably with the other injection modes and, in certain cases, outperforms them. The values obtained with SPME compare favorably to those reported for the analysis of polychlorinated biphenyls (PCBs), a typical environmental application [25]. Gorecki and Pawliszyn have evaluated SPME for the analysis of a variety of small hydrocarbons, including paraffins, olefins, and aromatics, in aqueous media and have reported similar % RSD values [26-27].

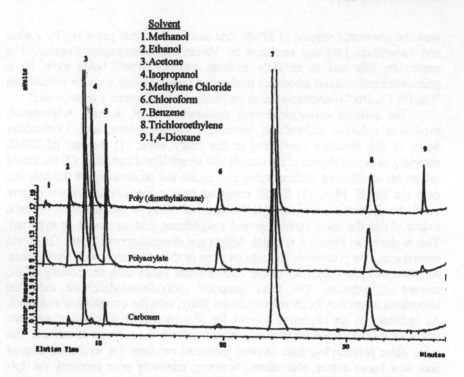

Figure 4.1 OVIs and common processing solvents evaluated with three different SPME fibers: poly(dimethylsiloxane), top chromatogram; Polyacrylate, middle; Carboxen, bottom. SPME conditions are headspace sampling, 15 minute adsorption, and 1 minute desorption into a direct injector at 200°C. GC analysis using DB-624 column, 75m x 0.53mm ID, 3 μm df (J&W Scientific) with FID at 260°C. Temperature program is 40°C initial (hold 35 min), 40°C/min, 200°C final (hold 5 min). From Ref. 19 (Courtesy of Roche, Inc., Nutley, NJ).

Table 4.1 System Precision (% Relative Standard Deviation)[a].

Solvent	Direct Injection	Static Headspace	SPME Sampling
Ethanol	1.7	3.2	3.7
Acetone	1.1	4.2	0.3
Isopropanol	1.0	4.3	1.9
Methylene Chloride	13.3	2.0	0.8
Chloroform	9.6	1.7	1.2
Benzene	3.9	2.1	0.9
Trichloroethylene	8.1	1.7	1.4
1,4 - Dioxane	1.2	4.4	2.7

[a] System precision compared for direct injection, headspace, and SPME using a (poly)dimethylsiloxane fiber.

Another important variable considered was the limit of detection (LOD). Such a value is calculated to be that amount of analyte yielding a response at two times the noise level. The LOD represents the smallest quantity that can be distinguished from baseline. Representative values are given in Table 4.2.

Table 4.2 Limit of Detection (Parts Per Billion)[a].

Solvent	Direct Injection	Static Headspace	SPME Sampling
Ethanol	n/a	100	5000
Acetone	150	10	200
Isopropanol	n/a	20	600
Methylene Chloride	500	20	60
Chloroform	1250	20	30
Benzene	200	0.4	2
Trichloroethylene	2490	20	20
1,4 - Dioxane	200	40	60

n/a = not available
[a] Representative analytes compared for direct injection, headspace, and SPME using a (poly)dimethyl-siloxane fiber.

The Limits of Quantitation (LOQ) represents the smallest quantity that can be measured and/or quantitated. To calculate the LOQ for these analytes, one would need to multiply the LOD values by five. In general, detection limits were poor for direct injection, whereas headspace and SPME showed drastic improvements in LOD values for the representative analytes considered here. However, SPME LOD values were not good for very volatile materials, presumably due to the incomplete partitioning into the fiber. In general, most solvents show respectable LOD/LOQ values with SPME and the requirements of the USP test <467> can certainly be met. In the event one needed to control a very toxic solvent, an alternate strategy might need to be considered.

Linearity from 10% to 150% of the expected concentration of the solvents was very favorable from all three techniques since the curves for all solvents were linear with correlation coefficients greater than 0.99 in each case. As linearity is really a test of the response range of the detector, it does indicate that utilizing SPME does not place one outside the working range of pharmaceutical solvent analysis.

Probably the largest test of the utility of SPME is shown by determination of method accuracy. To best explore the usefulness of this sampling technique compared to direct injection and static headspace sampling, spiked recovery measurements were made using two active substances, amitryptiline hydrochloride and pyridostigmine bromide, which vary greatly in their polarity and solubility characteristics. Table 4.3 summarizes the recovery values for amitryptiline hydrochloride under the conditions previously cited. Table 4.4 cites the method recoveries for pyridostigmine bromide drug substance. Because of the relative insolubility of pyridostigmine bromide in an aqueous medium, direct injection was not possible in this instance; therefore, only static headspace sampling and SPME were compared in this regard.

Table 4.3 Method Accuracy (% Recovery) for Amitryptiline hydrochloride drug substance[a]

Solvent	Direct Injection	Static Headspace	SPME Sampling
Ethanol	102.0	91.2	92.4
Acetone	103.3	85.3	97.6
Isopropanol	103.4	85.2	95.1
Methylene Chloride	89.0	84.1	78.5
Chloroform	115.0	82.6	63.9
Benzene	110.7	81.8	57.0
Trichloroethylene	115.9	81.9	53.0
1,4 - Dioxane	103.4	89.1	108.8

[a] Representative analytes compared for direct injection, headspace, and SPME on a (poly)dimethyl-siloxane fiber for amitryptiline hydrochloride drug substance.

Table 4.4 Method Accuracy (% Recovery) for Pyridostigmine Bromide Drug Substance[a].

Solvent	Direct Injection	Static Headspace	SPME Sampling
Ethanol	n/a	98.1	111.4
Acetone	n/a	97.3	101.2
Isopropanol	n/a	n/a	114.7
Methylene Chloride	n/a	92.0	90.7
Chloroform	n/a	92.1	93.1
Benzene	n/a	91.3	90.7
Trichloroethylene	n/a	91.5	92.0
1,4 - Dioxane	n/a	96.6	101.4

[a] Representative analytes compared for direct injection, headspace and SPME using a (poly)dimethylsiloxane fiber for pyridostigmine bromide drug substance.

These data indicate that SPME is virtually equivalent to static headspace sampling for water-insoluble drugs, with the added advantage of its simplicity, as evidenced from the data shown in Table 4.4. For soluble, active substances, such as amitryptiline hydrochloride, SPME falls short of headspace and direct injection for determining the OVIs. However, this does not imply that the technique did not hold merit for such analyses, as the aforementioned experiments were conducted on only a single fiber.

At the time the above work was performed, only a small number of fibers were available. Later work from Smith, Cafiero, and Scypinski utilized a Carbowax/Divinylbenzene (CW/DVB) fiber in the same automated instrument to explore applying SPME to a wider variety of substances and residual solvents [28]. In this rigorous study, a specific set of conditions were utilized to measure the utility of SPME in the headspace mode for analyzing residual solvents and OVIs in four drug substances of varying polarity and solubility. Using the same universal solvent method [7], fourteen common processing solvents and OVIs were separated, as shown in Figure 4.2. Of critical importance in this study was the system suitability criteria, which would be measured each time the method was run. In this regard, system suitability parameters are used not only to determine if the system is capable of performing the proposed analysis, but also as an ongoing check of the column performance. In other words, as the column ages, there is not a detrimental effect on the separation. Table 4.5 lists the fourteen solvents used in this work as well as the retention times, tailing factors, and resolution measurements. The retention times correspond with the peaks shown in Figure 4.2.

```
                          Solvents
              1.  Methanol          8.  Ethyl Acetate
              2.  Ethanol           9.  Chloroform
              3.  Acetone          10.  Benzene
              4.  Isopropanol      11.  Heptane
              5.  Acetonitrile     12.  Trichloroethylene
              6.  Methylene Chloride 13. 1,4-Dioxane
              7.  Hexane           14.  Toluene
```

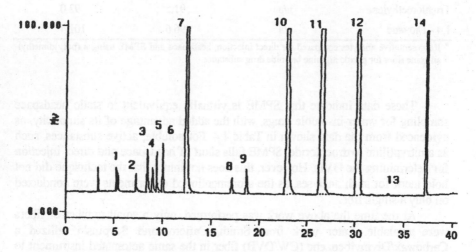

Figure 4.2 Separation of fourteen common processing solvents and OVIs with CW/DVB fiber. SPME conditions are headspace sampling from 0.8 mL sample volume, 15 minute adsorption time, 2 minute desorption time into a direct injector at 210°C. GC analysis using DB-624 column, 75 m x 0.53 mm ID, 3 μm df (J&W Scientific) with FID at 260°C. Temperature program is 40°C initial (hold 30 min), 8°C/min, 200°C final (hold 5 min). From Ref 28 (Courtesy of Roche, Inc., Nutley, NJ).

Clearly, the Carbowax/Divinylbenzene fiber was a logical choice for determining a wide variety of residual solvents and OVIs due to its high selectivity for the adsorption of a wide variety of solvents. To accurately determine the selectivity in a quantitative fashion, the relative degree of adsorption of the solvents was measured. This value, which provides a measure of the inherent selectivity of the fiber to effectively adsorb solvents, was calculated by measuring the standard peak area response ratio by SPME, versus direct injection. A value greater than unity implies that the SPME fiber is able to concentrate the material, meaning that the resulting response is greater than that seen when the analyte is injected directly onto the chromatographic system. These values are summarized in Table 4.6 and show that the CW/DVB fiber demonstrates excellent concentration of non-polar solvents and a majority of the OVIs. The response is acceptable, although the chemistry is still not ultimately optimal, i.e., the very polar solvents, such as the alcohols and 1,4-dioxane, show relative degrees of adsorption less than one. In this regard, the work described here found that the CW/DVB fiber was a good choice for the SPME determination of residual solvents and OVIs for a variety of drug substances. It is not uncommon that polar solvents can be problematic with respect to recovery and reproducibility in GC analysis. Various approaches to the analysis of polar solvents by SPME have been studied and summarized [29].

Table 4.5 SPME Evaluation Using the Carbowax/Divinylbenzene Fiber.

Solvent	Retention Time (Min)	Tailing Factor	Resolution[a]
Methanol	5.2	1.4	
Ethanol	7.0	3.2	14.2
Acetone	8.1	2.0	6.8
Isopropanol	8.7	1.7	3.0
Acetonitrile	9.0	2.2	2.1
Methylene Chloride	9.7	1.2	3.4
Hexane	12.1	1.1	10.7
Ethyl Acetate	16.6	1.4	15.6
Chloroform	18.1	1.1	4.5
Benzene	22.2	1.1	11.3
Heptane	25.6	1.0	7.8
Trichloroethylene	29.2	1.0	8.0
1,4-Dioxane	32.8	1.4	8.9
Toluene	37.1	1.0	19.4

[a] Resolution is calculated between each pair of adjacent peaks as described in USP 23 [6].

Table 4.6 Relative Degrees of Adsorption of Common Processing Solvents and OVIs Using the Carbowax/Divinylbenzene Fiber.

Solvent	Chemical Entity	Area $_{SPME}$/Area $_{DIRECT}$
Methanol	alcohol	0.1
Ethanol	alcohol	0.1
Isopropanol	alcohol	0.2
Acetone	ketone	0.3
Hexane	non-polar	666
Heptane	non-polar	1240
Toluene	non-polar	156
Methylene Chloride	OVI	10.6
Chloroform	OVI	24.5
Benzene	OVI	52.5
Trichloroethylene	OVI	122
1,4-Dioxane	OVI	0.2
Ethyl Acetate	other	1.9
Acetonitrile	other	0.3

The work done with the CW/DVB fiber was found to yield excellent system precision, as evidenced by the values summarized in Table 4.7. The data from this work showed that SPME can indeed produce excellent precision data on the order as headspace. Although direct injection showed even lower system precision than either SPME or headspace in some cases, a number of cases also occur where the system precision becomes prohibitive. This is especially true in the cases of the very non-polar solvents, such as hexane and heptane. Chloroform is also a problem. Santos, Galceran, and Fraisse applied SPME to the analysis of a variety of volatile organic compounds in water. They measured the time of adsorption to the fiber, linearity, detection limit, repeatability, and reproducibility. They also evaluated the effect of thickness of chemical coating on the fiber. Their findings agree closely with that reported by the Roche workers [30].

Table 4.7 System Precision for Common Processing Solvents and OVIs on the Carbowax/Divinylbenzene Fiber.

Solvent	%RSD$_{SPME}$	%RSD$_{HEADSPACE}$	%RSD$_{DIRECT}$
Methanol	2.5	2.0	0.6
Ethanol	4.0	3.2	0.9
Acetone	1.7	4.2	0.5
Isopropanol	2.5	4.3	2.3
Acetonitrile	1.2	-	3.6
Methylene Chloride	1.2	2.0	3.4
Hexane	3.1	2.1	40.4
Ethyl Acetate	1.8	-	2.3
Chloroform	1.3	1.7	13.6
Benzene	1.4	2.1	2.1
Heptane	2.8	-	39.0
Trichloroethylene	1.7	1.7	3.7
1,4-Dioxane	8.8	4.4	4.6
Toluene	1.4	2.2	7.0

This system and these conditions were applied to the analysis of four research phase drug substances of varying polarity and solubility. The identity of the drug substances is not as important as the data showing that SPME can be realistically used to determine the levels of solvents present in each material. For three of the four substances, SPME was directly compared with headspace sampling and with direct aqueous injection for the remaining substance. As can be seen from Table 4.8, SPME gives excellent results when compared to the techniques of choice that are currently in use today by most laboratories performing residual solvent testing of drug substances. In a similar fashion, SPME can be utilized for the analysis of volatile organic compounds in foods, which are an even more complex matrix than pharmaceutical dosage forms [31]. There is no doubt that SPME will assume its rightful place in the pharmaceutical analysis GC laboratory due to its usefulness, simplicity, and reduced cost when compared with companion techniques.

Table 4.8 Comparison of Actual Solvent Residue Levels as Determined by SPME and a Companion Technique.

Solvents Found	Amount by SPME	Amount by Headspace
Acetone[1]	nd	0.00005%
Isopropanol[1]	0.07%	0.06%
Methylene Chloride[1]	2.6 ppm	nd
Isopropanol[2]	0.11%	0.16%
Methanol[3]	nd	nd
Ethanol[3]	0.002%	0.003%
Acetone[3]	0.006%	0.001%
Methylene Chloride[3]	nd	nd
Ethanol[4]	0.023%	0.027%
Hexane[4]	0.00004%	-
Ethyl Acetate[4]	0.25%	0.26%
Heptane[4]	0.00002%	-
Toluene[4]	0.00001%	-

1 - Drug Substance A, Companion Technique - Headspace Sampling
2 - Drug Substance B, Companion Technique - Headspace Sampling
3 - Drug Substance C, Companion Technique - Headspace Sampling
4 - Drug Substance D, Companion Technique - Direct Injection
nd - None detected
(-) - Not evaluated

This typical and these conditions were applied to the
research areas that relate area of varying polarity and solubility. The volatility of
the drug substances is not as important as the ease showing that SPME can be
realistically used to determine the level of volatiles present in each material. For
sampling and with direct squeeze injection for
seen from Table 4.8. SPME gives excellent results when compared to the

SPME APPLIED TO OTHER PHARMACEUTICAL ANALYSIS AREAS AND DRUG DEVELOPMENT

The major portion of the work done in the various analytical chemistry groups and departments in the pharmaceutical industry does not involve GC, but more commonly employs high performance liquid chromatography (HPLC) as the technique of choice. As many pharmaceutical active ingredients and drug substances are nonvolatile, GC cannot perform measurements of their quality and potency unless special derivatization or other complex pretreatment is considered. In early stages of pharmaceutical development, it is most important to assimilate knowledge concerning the physicochemical characteristics of the drug substance, which is needed for formulation development. Such preformulation measurements are generally difficult, as a large quantity of substance is generally not available at this stage. Workers in England have reported a unique application of SPME as a means of measuring the octanol-water partition coefficient, which is a key parameter for assessing potential bioavailability of a compound [32]. Herbal medicines and their volatile degradation products have been analyzed successfully by SPME coupled with GC-MS [33]. Although at present, a reported application does not exist for the analysis of a pharmaceutical drug substance for potency by SPME or the assay of a dosage form by the aforementioned technique, the ability to do so has been reported. Chen and Pawliszyn have coupled SPME to HPLC and reported excellent results via the use of an interface [34]. Although their work was confined to environmental samples, there is no reason to believe that such an interface would not be practical for pharmaceutical applications, such as assay and content uniformity of tablet dosage forms. Such an interface might also be well adapted to the preparation of samples derived from biological fluids. Such analyses are tedious and are performed to determine trace concentrations of metabolites in such media as blood, serum, plasma, and urine.

Although a marked similarity exists to the work done for forensic applications (refer to Chapter 7), the focus of the work done for pharmaceutical development applications is to determine the pharmacokinetics and pharmacodynamics of new molecular entities and to support human clinical trials. Often, the absorption/distribution/metabolism/excretion (ADME) characteristics of a new drug are not thoroughly known or understood. However, the mission of the drug metabolism analytical laboratory is to recover as much of the drug in both its metabolized and unmetabolized states as possible. Many authors have reported results for determining drugs in biological fluids that would be acceptable if applied to drug metabolism and pharmacokinetic applications. The laboratory of Kumazawa at the Department of Legal Medicine in Japan has published numerous applications on the use of headspace SPME for determining various drugs in blood including anesthetics [35,36], cocaine [37], meperidine [38], phenothiazines [39], and tricyclic antidepressants [40]. Workers at the same institution have been successful in analyzing whole blood for hydrocarbons by employing SPME along with GC and chemical ionization selected ion monitoring mass spectrometry (CI-SIMMS) [41]. The coupling of CI-SIMMS with SPME resulted in detection limits 20 times more sensitive than conventional headspace sampling when employed for determining amphetamines in urine [42]. Lord and Pawliszyn showed in more recent work with the same chemical class that SPME is adaptable for determining a number of different substances in clinical or toxicological applications [43]. In a similar fashion, phencyclidine has been determined in blood and urine by headspace SPME by employing surface ionization detection (SID) [44]. GC was utilized in conjunction with SPME to quantify the narcotic methadone in urine down to a detection limit of 20 ng/mL [45]. Detection of ethanol in blood and urine samples at levels of 20 and 10 ng/mL, respectively, has been reported for SPME coupled with capillary GC [46]. Degel recently compared the merits of SPME with alternative methods of sample preparation, such as conventional Solid Phase Extraction (SPE) and various disk extraction devices. He concluded that SPME showed benefit in dedicated, selective extraction procedures with the advantages that SPME sampling is solvent-free, handling is easy, and extraction yields are good for substances that are applicable for analysis by the technique [47]. A Supelco application bulletin summarizes the analysis of drugs, alcohols, and organic solvents in biological fluid matrices and points out the relative advantages of SPME in this regard [48]. It will remain to be seen if this technique can concentrate the metabolic byproducts of drugs that have been ingested into and excreted from the body. Such metabolites are often difficult to quantify and preconcentration is often necessary.

CONCLUSIONS

This chapter has been a brief attempt to explore the role of SPME in the pharmaceutical analysis laboratory. Surely, this technique will find a rightful place in the pharmaceutical industry due to its ability to solve a variety of analytical problems. Any analysis can be divided into sample preparation, actual analysis, and data reduction and archival. The first step is often the most challenging. SPME offers simplicity and ease of use advantages, which should make it attractive for marketed product quality units. The continually increasing number of fiber types will enable it to be utilized in research areas as well. Finally, the availability of an HPLC interface will enable this exciting, new sample preparation tool to be coupled with the most common separation tool in the pharmaceutical analysis laboratory. We look forward to seeing many new developments in this area being reported in future literature.

REFERENCES

1. For general and current information on the Food and Drug Administration, the reader is encouraged to visit the FDA Website, http://www.fda.gov.
2. B Flickinger. ICH Guidelines: A primer on their practical implications. Pharmaceutical & Cosmetic Quality, 1:20-23, 1997.
3. JA Adamovics. Regulatory considerations for the chromatographer. In: JA Adamovics, ed. Chromatographic Analysis of Pharmaceuticals. New York: Marcel-Dekker, 1990, pp 3-20.
4. PK Hovsepian. Bulk drug substances and finished products. In: CM Riley, TW Rosanske, eds. Development and Validation of Analytical Methods. New York: Pergamon/Elsevier, 1996, pp 135-167.
5. J Pawliszyn. New directions in sample preparation for analysis of organic compounds. Trends Anal Chem 14: 113-122, 1995.
6. United States Pharmacopeia, 23rd edition, National Formulary, 18th edition Rockville, MD: USP, 1995, pp. 1746-1748.
7. L Clark, S Scypinski, and AM Smith. Proposed modifications to USP Method V for organic volatile impurities. Pharmacopeial Forum 19: 5067-5074, 1993.
8. LS Wigman, X Zhang, DO Fisher, and SW Walinsky. ICH Guideline: harmonization of residual solvents in pharmaceuticals. Pharmaceutical Technology 21: 102-108, 1997.
9. DW Foust and MS Bergren. Analysis of solvent residues in pharmaceutical bulk drugs by wall-coated open tubular gas chromatography. J Chromatogr 469: 161-173, 1989.
10. ID Smith and DG Waters. Determination of residual solvent levels in bulk pharmaceuticals by capillary gas chromatography. Analyst 116: 1327-1331, 1991.
11. BS Kersten. Drug matrix effect on the determination of residual solvents in bulk pharmaceuticals by wide bore capillary gas chromatography. J Chromatogr Sci 30: 115-119, 1992.
12. KJ Dennis, PA Josephs, and J Dokladalova. Proposed automated headspace method for organic volatile impurities <467> and other residual solvents. Pharmacopeial Forum 18: 2964-2972, 1992.
13. MV Russo. Static headspace gas chromatography of residual solvents in pharmaceutical products. Chromatographia 39: 645-648, 1994.
14. M DeSmet, K Roels, L Vanhoof, and W Lauwers. Automated headspace method for organic volatile impurities in drug substances dissolved in nonaqueous medium. Pharmacopeial Forum 21: 501-514, 1995.
15. CL Arthur and J Pawliszyn. Solid Phase Microextraction with thermal desorption using fused silica optical fibers. Anal Chem 62: 2145-2148, 1990.
16. Z Zhang, MJ Yang, and J Pawliszyn. Solid Phase Microextraction: A solvent-free alternative for sample preparation. Anal Chem 66: 844A-853A, 1994.
17. RF Mindrup. Solid Phase Microextraction simplifies preparation of forensic, pharmaceutical, and food and beverage samples. Chem NZ March: 21-23, March 1995.
18. J Pawliszyn. Solid Phase Microextraction: Theory and Practice. New York: Wiley-VCH, 1997, pp 141-191.

19. S Scypinski, AM Smith, LC Nelson, and SR Shaw. Application of Solid Phase Microextraction to determination of organic volatile impurities in pharmaceutical drug substances. Presented at the 1994 Pittsburgh Conference and Exposition on Analytical Chemistry and Applied Spectroscopy, Chicago, IL.

20. W Jennings. Analytical Gas Chromatography. Orlando, FL: Academic Press, 1995, pp 40-57.

21. RL Grob. Modern Practice of Gas Chromatography, Third Edition. New York: Wiley Interscience, 1995, pp 494-496.

22. BV Ioffe and AG Vitenberg. Headspace Analysis and Related Methods in Gas Chromatography. New York: Wiley Interscience, 1984, pp 67-97.

23. Supelco Corporation SPME Product Bulletin, 1997.

24. CL Arthur, LM Killam, KD Buchholz, J Pawliszyn, and JR Berg. Automation and optimization of Solid Phase Microextraction. Anal Chem 64: 1960-1966, 1992.

25. Y Liu, ML Lee, KJ Hageman, Y Yang, and SB Hawthorne. Solid Phase Microextraction of PAHs from aqueous samples using fibers coated with HPLC chemically bonded silica stationary phases. Anal Chem 69: 5001-5005, 1997.

26. T Gorecki and J Pawliszyn. Solid Phase Microextraction/Isothermal GC for rapid analysis of complex organic samples. J High Resolut Chromatogr 18: 161-166, 1995.

27. T Gorecki and J Pawliszyn. Sample introduction approaches for Solid Phase Microextraction/Rapid GC. Anal Chem 67: 3265-3274, 1995.

28. AM Smith, P Cafiero, S Scypinski. Solid Phase Microextraction (SPME): The use of SPME for the determination of residual solvents in pharmaceutical drug substances. Presented at the 19th International Symposium on Capillary Chromatography and Electrophoresis, Wintergreen, VA.

29. T Gorecki, P Martos, and J Pawliszyn. Strategies for the analysis of polar solvents in liquid matrixes. Anal Chem 70: 19-27, 1998.

30. FJ Santos, MT Galceran, and D Fraisse. Application of Solid Phase Microextraction to the analysis of volatile organic compounds in water. J Chromatogr A 742: 181-189, 1996.

31. BD Page and G Lacroix. Application of Solid Phase Microextraction to the Headspace gas chromatographic analysis of halogenated volatiles in selected foods. J Chromatogr 648: 199-211, 1993.

32. JR Dean, WR Tomlinson, V Makovskaya, R Cumming, M Hetheridge, and M Comber. Solid Phase Microextraction as a method for estimating the octanol-water partition coefficient. Anal Chem 68: 130-133, 1996.

33. J Czerwinski, B Zygmunt, and J Namiesnik. Headspace Solid Phase Microextraction for the GC-MS analysis of terpenoids in herb based formulations. Fresenius J Anal Chem 356: 80-83, 1996.

34. J Chen and J Pawliszyn. Solid Phase Microextraction coupled to high performance liquid chromatography. Anal Chem 67: 2530-2533, 1995.

35. T Kumazawa, X Lee, K Sato, H Seno, A Ishii, and O Suzuki. Detection of ten local anaesthetics in human blood using Solid Phase Microextraction (SPME) and capillary gas chromatography. Jpn J Forensic Toxicol 13: 182-188, 1995.

36. T Kumazawa, K Sato, H Seno, A Ishii, and O Suzuki. Extraction of local anaesthetics from human blood by direct immersion Solid Phase Microextraction (SPME). Chromatographia 43: 59-62, 1996.

37. T Kumazawa, K Watanabe, K Sato, H Seno, A Ishii, and O Suzuki. Detection of cocaine in human urine by Solid Phase Microextraction and capillary gas

chromatography with nitrogen-phosphorus detection. Jpn J Forensic Toxicol 13: 207-210, 1995.

38. H Seno, T Kumazawa, A Ishii, M Nishikawa, H Hattori, and O Suzuki. Detection of meperidine (pethidine) in human blood and urine by headspace Solid Phase Microextraction and gas chromatography. Jpn J Forensic Toxicol 13: 211-215, 1995.

39. H Seno, T Kumazawa, A Ishii, M Nishikawa, K Watanbe, H Hattori, and O Suzuki. Detection of some phenothiazines by headspace Solid Phase Microextraction and gas chromatography. Jpn J Forensic Toxicol 14: 30-34, 1996.

40. T Kumazawa, X Lee, M Tsai, H Seno, A Ishii, and K Sato. Simple extraction of tricyclic antidepressants in human urine by headspace Solid Phase Microextraction (SPME). Jpn J Forensic Toxicol 13: 25-30, 1995.

41. Y Iwasaki, M Yashiki, N Nagasawa, T Miyazaki, and T Kojima. Analysis of inflammable substances in blood using headspace Solid Phase Microextraction and chemical ionization selected ion monitoring. Jpn J Forensic Toxicol 13: 189-194, 1995.

42. M Yashiki, T Kojima, T Miyazaki, N Nagasawa, Y Iwasaki, and K Hara. Detection of amphetamines in urine using headspace Solid Phase Microextraction and chemical ionization selected ion monitoring. Forensic Sci Intl 76: 169-177, 1995.

43. H Lord and J Pawliszyn. Method optimization for the analysis of amphetamines in urine by Solid Phase Microextraction. Anal Chem 69: 3899-3906, 1997.

44. A Ishii, H Seno, T Kumazawa, K Watanabe, H Hattori, and O Suzuki. Simple extraction of phencyclidine from human body fluids by headspace Solid Phase Microextraction (SPME). Chromatographia 43: 331-333, 1996.

45. M Chiarotti and R Marsili. Gas chromatographic analysis of methadone in urine samples after Solid Phase Microextraction. J Microcol Sep 6: 577-580, 1994.

46. T Kumazawa, H Seno, X Lee, A Ishii, O Suzuki, and K Sato. Detection of ethanol in human body fluids by headspace Solid Phase Microextraction (SPME)/capillary gas chromatography. Chromatographia 43: 393-397, 1996.

47. F Degel. Comparison of new Solid Phase Extraction methods for chromatographic identification of drugs in clinical toxicological Analysis. Clin Biochem 29: 529-540, 1996.

48. Solid Phase Microextraction/capillary GC analysis of drugs, alcohols, and organic solvents in biological fluids. Supelco Applications Bulletin 901, 1997.

5

Environmental Applications

Brian MacGillivray[†]
Water Technology International Corporation, Burlington, Ontario, Canada

INTRODUCTION

The modern environmental industry demands rapid, sensitive, low cost chemical analyses for effluent monitoring, contaminated site remediation, and development of regulatory guidelines. Over the past decade, public sector and private laboratories alike have striven to reduce costs by simplifying and automating routine analytical procedures. Sample preparation steps can be time consuming, as is demonstrated by many of the current organic solvent extraction methods. They often require expensive instrumentation, as in the case of *purge and trap* analysis for volatile organics in aqueous matrices. Consequently, significant research has been invested in improving current methods as well as in alternative methods development.

Solid Phase Microextraction (SPME) as a sample preparation technique has several apparent advantages over conventional methods for specific environmental applications. It requires no solvents and permits sample transfer and analysis with little or no modifications to existing chromatographic equipment. As a result, implementation costs can be kept low. Key advantages of utilizing SPME for environmental work include its simplicity, amenability to automation, and suitability for field and on-site applications due to its portability. This section is concerned

---

[†]Deceased.

with specific applications of SPME to environmental chemical parameters, particularly organic compounds. Consideration will be given to the performance of SPME against traditional methods for extraction of organics, its advantages and limitations, and the recent status of environmental SPME research with respect to alternate environmental applications.

TRADITIONAL ENVIRONMENTAL METHODS

A comprehensive review of current environmental sample preparation techniques is beyond the scope of this chapter. It is, however, important that the principles of pertinent methods be understood by the SPME user who will ultimately require a comparison against benchmark methods in the development and validation of SPME techniques. Although it is not the sole source of accepted analytical methods, the U.S. Environmental Protection Agency (EPA) has published established reference procedures widely used in North America and elsewhere. These methods address the analysis of organic constituents in drinking waters [1], in wastewaters [2], in solid wastes such as soils and sediments [3], and in ambient air [4]. A widely referenced publication of standard methods [5] describes analyses in various matrices. Other organizations, such as the Association of Official Analytical Chemists (AOAC) and American Society for Testing and Materials (ASTM), publish equivalent methods to those of the EPA. Newly developed and modified sample preparation techniques are continually being proposed for acceptance by regulatory bodies, but may not yet have been accepted for inclusion into standard methods.

Waters and Wastewaters

In the broad sense, water samples can consist of drinking water, groundwater, wastewater, or wet sludge, provided the solids content is less than a few percent. Categorically, organic compounds are divided into either volatile or semivolatile/nonvolatile groups; the boundary between the two groups remains somewhat arbitrary, although volatiles are often defined as having boiling points below 200°C. In practice, sample preparation methods in use in North America for extracting semivolatiles from aqueous samples make use of some form of liquid-liquid or liquid-solid extraction. A typical liquid-liquid extraction employs a separatory funnel and organic solvent to extract parameters from aqueous samples that can then be introduced to either gas or liquid chromatographic systems for analysis. Separatory funnel extractions are inappropriate for some liquid samples due to excessive emulsion formation that prevents post-extraction separation of the organic and aqueous phases; the emulsions are often a consequence of turbulent mixing. Continuous liquid-liquid extraction [3] offers a solution by using a distillation apparatus to gently cycle the extraction solvent repeatedly through a water sample. It is a time-consuming process, requiring 18-24 hours per extraction, and is therefore not normally feasible for processing large numbers of samples simultaneously. Consequently, continuous liquid-liquid extractions tend to be avoided where separatory funnel extraction will suffice.

Another alternative for extracting aqueous samples is Solid Phase Extraction (SPE), a form of liquid-solid extraction. The procedure involves isolation of target analytes by passing an aqueous sample through a coated membrane or particulate bed mounted in a filtration apparatus. Specific functional groups in the coated support have an affinity for the analytes of interest, retaining them effectively in the solid phase. Upon completion of the filtration, the analytes are quantitatively eluted from the solid with an appropriate solvent. As with SPME, solid phase coatings can be designed for specific applications. Typically, coatings of C_{18} are used for non-polar analytes. The method has been used for extracting organochlorine pesticides, phthalate esters, other selected semivolatile organics, and metals in solution.

For volatile organics in water and wastewater, one of the most widely used methods is dynamic headspace analysis, commonly known as Purge and Trap (Figure 5.1). Volatiles are purged from a liquid sample by bubbling an inert gas (such as helium) through a sample then they are swept onto a solid sorbent trap where they are retained and concentrated. The organics are then thermally desorbed from the trap using additional helium and directed to the head of a gas chromatograph for separation and measurement. Static, or equilibrium, headspace is a variation on the method in which an aliquot of headspace over a heated, enclosed sample is taken at equilibrium. As with Purge and Trap, analytes are transferred to a gas chromatograph (GC) for quantitation. Optional methods [3] tailored for more specific applications include vacuum distillation (Method 5032) and closed-loop stripping (Method 5035).

Solids

Solid sample extraction methods are applicable to soils, sediments, dewatered and dried sludges, clays, and other waste solids. Most samples for semivolatiles and nonvolatiles have traditionally been prepared by soxhlet extraction. The recent development of automated soxhlet extraction, which uses a heated solvent for extraction rather than solvent at ambient temperature, has allowed the reduction of both extraction duration and solvent usage while achieving comparable end results.

Supercritical fluid extraction (SFE) has advanced significantly in recent years, having become an accepted method for the extraction of PAH and other semivolatiles from solid samples. Accelerated solvent extraction (ASE) uses heat and pressure to extract organics with solvent and, like the automated soxhlet procedure, aims to save on preparation time and solvent usage. Current research continues into the use of microwaves to assist in separation of target analytes from solid matrices; and microwave techniques may yet be incorporated into regulatory methods for organics in North America.

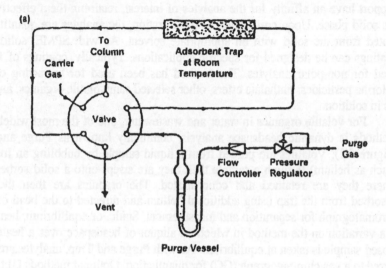

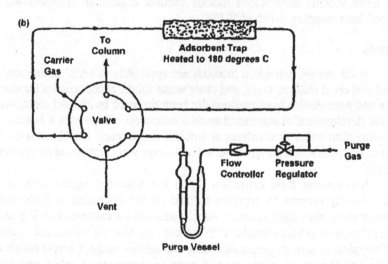

Figure 5.1 Flow diagram of a Purge and Trap concentrator in (a) purge mode and (b) desorb mode. (From Ref. 53.)

Purge and Trap Method 5030 for solid wastes [3] describes two methods for solids in which volatile components are being measured. Low-level samples, defined as containing less than 1 μg/g of analyte, are made into an aqueous slurry, spiked with surrogate compounds to monitor recovery, and introduced to a Purge and Trap system. Solids containing higher levels of contaminants are extracted with an appropriate solvent, such as methanol. The extract is then either introduced by direct injection to a gas chromatograph or spiked into water for Purge and Trap/GC analysis.

Air

Various approaches have been taken to air sampling in matrices ranging from ambient air containing trace contaminants, to source gas collected from an industrial discharge stack. Ambient air sampling typically involves either analyte extraction from an air stream that passes through a bed of solid sorbent material, or acquisition of whole air samples in specifically designed collection vessels. The sorbent beds are contained in cylindrical tubes of metal or glass that can be connected to sample trains that meter the flows through them. Volatiles collected in sorbent tubes are solvent eluted or thermally desorbed with an inert gas prior to introduction to an analytical system, usually a GC. Whole air samples are delivered from specially constructed sampling bags or metal canisters to cryogenic concentrators that quantitatively introduce an aliquot of the sampled air to the GC. Semivolatiles in the vapor phase may be collected by flowing air through sorbent beds packed with polyurethane foam and/or polymeric resins designed to adsorb them. Soxhlet apparatus is used to extract target analytes from the sampling beds, followed by GC analysis with the appropriate detector. Industrial stack sampling and analysis is a rather specialized discipline, and is described in volatile organic sampling train (VOST) methods 0030/5041 [3].

These methods represent the most widely used techniques; certainly others exist, which in some cases are modified versions of present regulatory methods. For less rigorous treatment or for rapid screening of air samples, several real-time instruments exploiting infrared absorption or photoionization potential are available. The U.S. EPA has assembled a compendium of methods for toxic organics in ambient air [4] that serves as a useful reference.

SPME APPLICATIONS

Practical Considerations

Matrix

Applications of SPME to several matrices have been detailed in the literature, including liquids, solids and air. Environmental samples by their nature are often highly contaminated, and in many instances their characteristics are unknown. Since much of modern laboratory instrumentation is designed for trace analysis, the same care must be taken as with conventional sample extraction to avoid fiber and/or equipment contamination. Although it is seldom possible for all matrix effects to be known, compounds with high K_{fs} values are more difficult to desorb from an SPME fiber than those that partition relatively less into the fiber.

Carryover of volatile substituted benzenes [6] and PAHs [7] due to incomplete desorption, also referred to as "memory effects", has been reported with poly(dimethylsiloxane) (PDMS) coated fibers. Caution is advised, particularly when encountering unknown samples that will be extracted by direct immersion of the fiber in a liquid. In addition to monitoring for carryover, the potential effects of direct exposure of dirty samples to the surface of the fiber, and therefore on its subsequent performance, must be investigated and characterized for each application. Moreover, samples such as sludges may be sufficiently viscous as to hinder mechanical sample mixing efficiency, regardless of the agitation technique used. Additionally, the potential exists for contamination of injection liners or flow blockages due to septum fragments being transported into the injector by the SPME needle.

Logical alternatives for handling complex and highly concentrated samples include reducing extraction time, diluting the sample, and employing a thinner stationary phase. All of these strategies reduce the quantity of analyte sorbed by the fiber. Some selectivity can be introduced by using a headspace approach, most notably where interest is in compounds that have a combination of a sufficient Henry's Constant and fiber/sample partition ratio. Headspace SPME conveniently eliminates concerns of absorbing non-volatile, high K_{fs} components, which must be baked out of a GC system, and helps to prevent fiber/injector fouling. The addition of salt, (i.e., sodium chloride or sodium sulfate), to aqueous samples can reduce the solubility of moderately polar compounds such as benzene. In turn, the resulting shift in equilibrium in favor of sorption to the solid phase enhances sensitivity of both the direct immersion and headspace techniques [8].

When sampling air or gas, handling the fiber to maintain the integrity of a sample is of particular importance. Volatiles having low fiber/gas partition coefficients may be released from the fiber if a significant period of time lapses between extracting a gas sample and introducing it into the desorption mechanism. A thicker film releases absorbed analytes more slowly and allows more time for fiber transfer to the analytical system without adversely affecting results. Determinations using BTEX sampled from water [9] found that a fiber coated with PDMS (56 μm thickness) retained benzene for 2 minutes when held to outside air between the extraction step and desorption. Similarly, xylenes were retained for 5 minutes before significant reductions in concentrations were observed. An expanded study [10] examined the effects of temperature, storage time, and sealing the retracted fiber in the syringe needle barrel with a simple GC septum. The outcome is summarized in Table 5.1. There are clearly implications when considering that sampling in field conditions if SPME devices are far removed from analytical instrumentation; storage and transport of fibers does not eliminate the potential for analyte losses or contamination while in transit, and should be considered carefully. In such instances, the use of portable or field instrumentation expels these concerns.

Soils and other solids commonly present the difficult challenge of obtaining analytical precision that is deemed acceptable. Frequently, this is more a function of matrix homogeneity and the sampling process, rather than of the analyst or the method itself. It is not only SPME users that have to contend with troublesome solid samples. Several sample preparation options, including those derived from regulatory methods, can, and do, deliver quite varied results with these materials. Matrix considerations of solids will be further addressed in subsequent sections of this chapter.

Fiber Chemistries and Configurations

Stationary phase coatings for SPME fibers continue to evolve for specific classes of organic chemical compounds in much the same manner as capillary GC column phases. As with traditional liquid-liquid extractions, there is a balance between specificity and sensitivity. Environmental chemical contaminants have historically been classed, somewhat ambiguously, into two categories; polar and non-polar. Similarly, most organics are referred to as either semivolatile or volatile. The most common fiber in use to date has been the non-polar PDMS, simply due to the vast number of non-polar environmental contaminants that are regulated. It is also versatile enough to be used with some moderately polar analytes such as chlorophenols. Originally, PDMS fibers were hand-made by experimenters from coated fused silica optical fibers; now they are conveniently obtained in syringe-ready kits supplied commercially.

Table 5.1 Percentage of Analyte Retained in Fiber Coating Under Various Storage Conditions.

Compound	25°C Uncapped (%) 2 min	25°C Capped (%) 60 min	25°C Capped (%) 30 min	25°C Capped (%) 60 min	5°C Capped (%) 30 min	5°C Capped (%) 60 min	-70°C Capped (%) 60 min	-70°C Capped (%) 24 hr	-70°C Capped (%) 48 hr
Chloroform	97	51	70	62	93	89	95	50	30
1,1,1-trichloroethane	95	49	67	49	93	91	92	76	45
Carbon tetrachloride	97	45	73	70	95	93	95	79	66
Benzene	94	69	89	81	94	92	97	83	82
Toluene	96	77	86	89	95	95	94	92	85
Tetrachloroethylene	96	79	85	88	95	94	99	97	90
1,1,2,2-Tetrachloroethane	98	70	90	80	94	92	98	95	93

Source: Ref. 10.

Volatiles tend to be best extracted with a thicker (100 μm film) PDMS stationary phase. Partitioning coefficients of non-polar, semivolatile components are large; therefore, a thin-film PDMS phase (30 μm or 7 μm) will not only provide the required sensitivity, but will also reduce fiber carryover between samples. For more polar contaminants, such as phenols, a polyacrylate stationary phase is more appropriate. As development has moved forward, polymers and carbon-based adsorbent materials have been combined to offer increased specificity for volatiles. Poly(dimethylsiloxane) and divinylbenzene have been combined in a coating designed to extract polar, volatile compounds. Often, more than one type of coating can be used to extract a specific contaminant; as a result, the most appropriate stationary phase can ultimately be determined by the judgment and experience of the analyst.

When considering the choice of fiber, care must be taken to observe the upper temperature limit of the coated solid phase. If recommended GC injector temperatures are exceeded, phase decomposition can produce bleed, and subsequent interfering peaks. In view of the fact that fiber coatings vary in durability and robustness, routine checks using a stable calibration standard ensure consistent performance and data that is comparable between tests. Upon startup, it is prudent to experimentally establish equilibration times, sample carryover between analyses, precision, and linear range for each new solid phase.

Automation

Automation significantly enhances the efficiency and reproducibility of a laboratory procedure; however, not all procedures are amenable to being automated, and for small workloads the ability to automate is less critical. There are several key considerations when implementing automation of SPME for environmental sample analysis. The automated system must be capable of maintaining the fiber sheathed in the syringe barrel while both entering and exiting the sample vial through a septum. It must be capable of controlling the sampling time to allow for differing equilibration times, and it should control the vertical position of the fiber inside the vial. In addition, the automated system should coordinate with the analytical instrument cycle time to control when the absorption begins for the next sample; consequently, minimizing the fiber exposure before desorption. A fiber exposed to the laboratory atmosphere for several minutes could either lose volatile components that have been sampled or absorb contaminants prior to sampling, depending on the sequence of the exposure. Accordingly, exposure of the sheathed fiber to laboratory air within the analytical sequence is best minimized.

In non-automated systems, it is common to elicit mixing of a liquid sample by means of a magnetic stirbar or a similar device. No commercially available automated SPME sampler using a stirbar-type of mechanism is available at the time of this writing. Rather, immersion of the fiber into the aqueous phase and subsequent high-frequency vibration of the fiber by an external mechanism [11] achieves sample agitation. The analysis of semivolatiles in relatively clean samples, having neither excessive viscosity nor solids content, is an ideal application for this type of system. In considering a headspace configuration, the vibrating fiber does not agitate the aqueous phase. If the gas/liquid partition coefficient (effectively the Henry's Law coefficient) of an analyte is high and the fiber/gas partition coefficient is low, equilibration times should not be significantly longer with a static aqueous phase than in the stirred case. This is true, provided the headspace and liquid are in equilibrium when the fiber exposure commences [12]. Under these conditions the absolute amount of analyte absorbed by the fiber will be small in comparison to the amount of analyte in the vapor phase. Therefore, the concentration in the liquid phase remains almost constant during the extraction. Because the rate of attainment of equilibrium is then essentially dependent upon diffusion of analyte through the gas phase, it remains relatively rapid. Most volatile analytes meet these criteria, and so are amenable to static headspace analysis provided gas/liquid equilibrium conditions can be consistently assured prior to extraction.

Those interested in customizing existing automatic samplers for headspace SPME should consider that many samplers accommodate sample vials having a capacity of 2 mL or less. Given that the size of the extraction vessel may limit the ability of the analyst to reproducibly introduce a non-homogeneous sample as well as stir it efficiently, vials of approximately 10 mL capacity or more are preferable in some situations. The volume of sample vials is particularly critical to headspace analysis to allow ample room for the fiber above the sample. As with manual analyses, efficient mixing is desirable for both rapid attainment of equilibrium and reproducibility. As samples may be inhomogeneous or high in solids content, it is necessary to establish the range of variability for each matrix for a given configuration of autosampler, to ensure that it meets the requirements of the user.

For the relatively specialized process of air sampling and analysis, SPME automation has not advanced to the stage of commercial production, but could do so in the near future if demand materializes.

Sensitivity

This is perhaps the topic most often discussed in environmental chemistry, and indeed its relevance drives the continued quest for lower detection limits. The logical aim is to first consider the sensitivity requirements of individual situations (so-called *data quality objectives*) and aim to reasonably meet them without unduly expending time and effort to far exceed them. In many instances, an inexpensive detector and a simple GC temperature program will deliver the desired results. In determining method detection limits statistically, precision can have a greater effect on the calculated result than the potential sensitivity of the instrument or method. It is therefore critical to replicate the procedure as closely as possible, both for samples and calibration mixtures. Extraction time should be consistent, and volume of sample sufficient to gain a representative and reproducible aliquot.

The response of a compound to a specific detector, its solubility, the cleanliness of the analytical system, fiber chemistry, and film thickness are key factors affecting detection limits. We are reminded, for example, that the amount of analyte absorbed is proportional to the partition coefficient, and assuming a negligible fiber core diameter, is proportional to the square of the phase thickness. In many cases, the optimum fiber type and phase thickness can be determined in advance of the analytical set-up if the partition coefficient is known or if it can be reasonably estimated. Because of numerous environmental analytes and method variations, sensitivity will be treated in more detail by another section.

Volatile Organic Compounds

It is well established that volatiles analysis of aqueous media for volatile organic compounds is amenable to either direct sampling with the fiber immersed in sample or headspace sampling. Recall the equations for the amount of analyte, n_f, absorbed by the fiber:

a) directly from water:

$$n_f = \frac{K_{fs}V_fV_sC_o}{K_{fs}V_f + V_s}$$ (5.1)

where C_o is the initial concentration of analyte in solution; V_f and V_s are the volumes of the fiber coating and the aqueous phase, respectively, and K_{fs} is the partition coefficient between the fiber coating and the aqueous phase;

b) from headspace over a water sample:

$$n = \frac{C_oV_fV_sK_{fs}}{K_{fs}V_f + K_{hs}V_h + V_s}$$ (5.2)

where V_h is the volume of headspace and K_{hs} is the partition coefficient between the fiber coating and the gas phase.

If the headspace volume, V_h, in equation 5.2 is much less than V_s, and K_{hs} is small, the effect of the headspace capacity term $K_{hs}V_s$ is minimized and the sensitivity of the headspace method approaches that of direct immersion sampling. The headspace technique is desirable in especially complex or contaminated matrices in which fouling of the fiber, as a result of immersion, is a concern. Headspace sampling also provides the analyst with a degree of selectivity with respect to limiting the content of unwanted semivolatiles absorbed by the coating [12]. As a case in point, a poly(dimethylsiloxane)-coated fiber used in the headspace SPME mode permits extraction of substituted benzenes from an aqueous petrochemical sample while limiting heavier polynuclear aromatic hydrocarbons (PAHs) that would be absorbed during direct immersion. For relatively clean samples, such as drinking water or most groundwaters and surface waters, the technique may be applied with either direct immersion of the fiber in the sample or with the headspace configuration. The addition of surrogate compounds to monitor extraction efficiency and recoveries provides the same advantages as in traditional methods for volatiles.

Nonpolar Volatiles in Aqueous Samples

The use of SPME for the analyses of EPA lists of volatiles in drinking waters and wastewaters has been clearly demonstrated [13,14]. These standard lists, such as those given in EPA Methods 524.2 for drinking waters [1] and 624 for wastewaters [2], are comprised chiefly of substituted benzene compounds as well as a host of halogenated aromatics and aliphatics. Since most regulatory listed volatiles are non-polar and are observed below 10 µg/L, the 100 µm PDMS fiber has been the most useful for this application, so far.

The method of choice for analysis is gas chromatography, and in virtually all current development and analysis for volatiles, capillary GC columns are used. Due to the wide use of Purge and Trap and static headspace techniques for volatiles analysis, several GC methods and column options are detailed in both the scientific literature and in laboratory trade literature. Furthermore, most manufacturers of custom GC columns for volatiles analysis can supply recommended temperature programs and gas flows for specific target lists. It is, however, necessary to consider the process of fiber desorption and its effect on peak bandwidth. Depending on the conditions specified in a typical isothermal GC injector, deposition of analyte onto the head of the GC column can occur over a longer period than introduction by a direct solvent injection, broadening chromatographic peaks. Peak bandwidth is dependent on several interrelated variables, one being the speed of desorption, which is in turn affected by injector temperature and volatility of the analyte. Rapid heating of the fiber is preferred, which implies setting the injector temperature to the highest practical value and maintaining it isothermally.

The type and volume of the GC injection sleeve can have a marked effect on peak width, especially on the lower boiling volatiles. Minimizing the internal volume, or dead volume, helps to maintain sharp peaks, as previously demonstrated in Figure 3.8. Splitting the flow in a split/splitless injector is ordinarily not necessary due to the absence of a solvent peak. The optimum position of the fiber in the injection liner during desorption is in close proximity to the liner/column interface but at a sufficient distance (1-2 mm) from any obstructions to prevent the possibility of physical damage to the fiber. Inasmuch as this adjustment is instrument-specific and performance may also be affected by temperature gradients within the desorption zone, it is best determined experimentally.

Refocusing the analytes into a narrow band at the head of the column is a tool well known by chromatographers and, when utilized properly, can counter the effects of slow desorption or excess injector dead volume. Increasing the column phase thickness and lowering the initial oven temperature are the key methods of refocusing the analyte band to enhance peak resolution. The GC oven can be adjusted to sub-ambient temperatures with the aid of a cryogen, such as liquid nitrogen or pressurized carbon dioxide; however, the obvious complexity and additional costs of cryogenic cooling suggests that thick film columns are preferable if they can resolve the highly volatile compounds. In going to a thicker film, such as a 1.5 μm thickness or greater, column efficiency is reduced and peak broadening of the later eluting compounds will increase. A higher temperature stage will be necessary at the end of the GC run to elute the less volatile compounds, however this does not typically present a significant problem. The tangible results of column refocusing are improved reproducibility and enhanced chromatographic separation of the more volatile, early eluting compounds (such as chloromethane and bromomethane), which have a tendency to coelute.

For the gas chromatographic separation, the specific situation and requirements determine the combination of variables such as column diameter, flow rate, oven program, and film thickness. For example, a 30 m x 0.25 mm I.D. fused silica VOCOL® column with 1.5 μm film used an initial column temperature of -15°C and a temperature ramp to 142°C to separate the 31 compounds in EPA Method 624. The concentration was 75 μg/L in water and extraction time was 5.0 minutes. The fiber was 100 μm PDMS desorbed at 200°C for 3.0 minutes with a GC run time of 24 minutes [15]. If the column were to have a thin (0.5 μm) film thickness, an initial column temperature as low as -40°C would be required to achieve similar performance. The same list of compounds has been separated after desorption from a 100 μm PDMS fiber using two columns types with varied dimensions [14], summarized in Table 5.2. The short analysis times of 3.5 to 4.0 minutes compromises the separation of the light gases at the front end of the chromatogram. Conversely, use of a 60 meter column resolves the early peaks, but results in increased peak broadening of later eluting peaks and adds to total time of the analysis. Figure 5.2 illustrates separation of 60 volatile organics after SPME in 50 minutes with the extended column. Table 5.3 identifies these compounds. There is potential for coelution of vinyl chloride with methanol, provided an MS detector is available so small amounts of methanol will not interfere with quantitative analysis.

Table 5.2 Analysis Times for VOCs on Two GC Columns.

	Analysis Time (min)	
Column Dimensions	SPB-1 Column	VOCOL Column
60 m x 0.25 mm I.D. x 1.5 μm	28.5	29.5
20 m x 0.2 mm I.D. x 1.2 μm	11.5	12
10 m x 0.2 mm I.D. x 1.2 μm	5.5	6
5 m x 0.15 mm I.D. x 0.9 μm	4.5	5
3 m x 0.1 mm I.D. x 0.6 μm	3.5	4

Source: Ref. 14.

The choice of detector(s), whether mass spectrometric (MS) or dedicated detectors, such as the electron capture detector (ECD) or flame ionization detector (FID), will affect both the sensitivity that can be attained and the degree of chromatographic resolution required. By using a MS detector, coeluting peaks can be resolved using extracted single ion chromatograms if the coeluting compounds have distinguishable ion signals. If either FID or ECD is the chosen detector, a dual column/dual detector analysis using columns of differing polarities is recommended to provide compound confirmation. Detection limits at part-per-trillion concentrations have been reported [16] using headspace SPME with the highly sensitive ion trap MS detector. Table 5.4 indicates part-per-trillion (ng/kg) sensitivity of six common volatiles in wastewater, aqueous sewage sludge, and sand.

To achieve such low detection levels, the addition of salt to the aqueous matrix may be necessary. A series of instrument responses was collected for EPA Method 624 target analytes in water, reproduced in Table 5.5. Note that the addition of salt appears to lower the sensitivity of some compounds, while raising it for others, further indicating the need to optimize conditions for specific components and target groups prior to establishing routine procedures. Direct immersion analyses are best done with minimum headspace in the vials, and as alluded to earlier, headspace SPME is recommended for wastewaters and sludges.

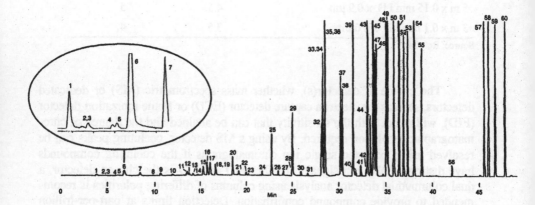

Figure 5.2 U.S. EPA Method 524.2 volatile analytes using SPME. 50 ppb of each analyte (listed in Table 5.3) in drinking water. Headspace extraction using PDMS fiber, 100 μm film, for 5 min. Desorption for 6 min, 220°C, splitless injection at 220°C. VOCOL column, 60 m x 0.25 mm, I.D., 1.5μm film. Column oven at 35°C for 4 min. Programmed to 200°C at 4°/min. Helium carrier, 2 mL/min. MS detection (m/z = 35-260, 0.6 sec/scan). (Courtesy of Supelco, Bellefonte, PA.)

Table 5.3 Analytes in U.S. EPA Method 524.2.

Components		
1. Dichlorofluoromethane	21. 1,2 Dichloropropane	41. 1,1,2,2-Tetrachloroethane
2. Chloromethane	22. Bromodichloromethane	42. 1,2,3-Trichloropropane
3. Vinyl chloride	23. Dibromomethane	43. *n*-Propylbenzene
4. Bromomethane	24. *cis*-1,3-Dichloropropene	44. Bromobenzene
5. Chloroethane	25. Toluene	45. 1,3,5-Trimethylbenzene
6. Trichlorofluoromethane	26. *trans*-1,3-Trichloropropene	46. 2-Chlorotoluene
7. 1,1-Dichloroethylene	27. 1,1,2-Trichloroethane	47. 4-Chlorotoluene
8. Methylene chloride	28. 1,3-Dichloropropane	48. *tert*-Butylbenzene
9. *trans*-1,2-Dichloroethylene	29. Tetrachloroethylene	49. 1,2,4-Trimethylbenzene
10. 1,1-Dichloroethane	30. Dibromochloromethane	50. *sec*-Butylbenzene
11. 2,2-Dichloropropane	31. 1,2-Dibromoethane	51. Isopropyltoluene
12. *cis*-1,2-Dichloroethylene	32. Chlorobenzene	52. 1,3-Dichlorobenzene
13. Chloroform	33. 1,1,1,2-Tetrachloroethane	53. 1,4-Dichlorobenzene
14. Bromochloromethane	34. Ethylbenzene	54. *n*-Butylbenzene
15. 1,1,1-Trichloroethane	35. *m*-Xylene	55. 1,2-Dichlorobenzene
16. 1,1-Dichloropropene	36. *p*-Xylene	56. 1,2-dibromo-3-chloropropane
17. Carbon tetrachloride	37. *o*-Xylene	57. 1,2,4-Trichlorobenzene
18. 1,2-Dichloroethane	38. Styrene	58. Hexachlorobutadiene
19. Benzene	39. Isopropylbenzene	59. Naphthalene
20. Trichloroethylene	40. Bromoform	60. 1,2,3-Trichlorobenzene

Table 5.4 Precision and Limit of Detection (LOD) for Volatile Organic Compounds in Three Matrices at 5 ppb Concentration Level.

	Matrix					
	Wastewater		Aqueous Sludge		Sand	
	Prec.	LOD[a]	Prec.	LOD[a]	Prec.	LOD[a]
Compound	(%)	(ppt)	(%)	(ppt)	(%)	(ppt)
1,1-Dichloroethane	14	400	14	450	4	80
Chloroform	5	6	9	150	2	40
Carbon tetrachloride	6	20	3	70	4	20
Trichloroethane	7	110	7	550	6	70
Dibromochloromethane	3	210	7	110	3	3
Chlorobenzene	14	2	6	30	4	1

[a]LOD is calculated from mass spectral data assuming a signal-to-noise ratio of 3.
Source: Ref. 16.

Desorption temperatures of 200-230°C and durations of 3-4 minutes are normally sufficient for volatiles when utilizing the 100 μm PDMS fiber. Though the injector is ideally operated isothermally, temperature programming can be used if there are concerns of thermally labile components decomposing during desorption. Desorption temperature for the PDMS fiber can be safely set to 250°C without causing deterioration of the stationary phase. Elevated desorption temperatures are necessary if an unknown sample is found to contribute unwanted contaminants or if analyte levels are sufficiently high as to result in significant carryover. Alternatively, extraction times may be shortened to reduce carryover provided the times are kept consistent for both instrument calibration and analysis. The increase in detection limits experienced by reducing absorption times is obviously not a concern if sample concentrations are sufficiently high. Many samples actually require dilution to place them within the linear confines of the analytical method. Analysis of highly contaminated samples requires a prolonged bakeout period of the GC column to prevent late-eluting peaks from contaminating subsequent analyses.

Table 5.5 Sampling Method and Salt Effect.

	Headspace		Immersion	
	Salt[a]	No Salt	Salt[a]	No Salt
Benzene	5400	4200	17500	4300
Bromodichloromethane	2500	1100	4600	900
Bromoform	3600	1200	3600	1000
Carbon tetrachloride	2100	4200	7800	6200
Chloroethane	0	0	600	0
2-Chloroethylvinyl ether	300	0	1100	0
Chloroform	1500	1000	1900	1000
1,3-Dichlorobenzene	41100	39200	15300	15700
1,1-Dichloroethene	200	500	2900	900
1,2-Dichloropropane	1900	800	3800	900
cis-1,3-Dichloropropene	4200	1400	7100	1600
Ethylbenzene	39900	51500	34000	31500
Methylene chloride	600	200	1600	200
1,1,2,2-Tetrachloroethane	10400	2300	8000	2000
Tetrachloroethane	4800	9800	6300	6600
Toluene	15500	14500	26800	13800
Trichloroethene	2200	2600	5700	2800
Trichlorofluoromethane	400	1100	7200	4300
Vinyl chloride	0	>0	1200	<100

[a]Sodium chloride, saturated solution. Analytes listed in U.S. EPA Method 624.
Values = area counts (to nearest 100).
Source: Ref. 8.

The importance of sample agitation has been discussed on several occasions in the literature. A commercial manufacturer of SPME fibers has indicated that poor stirring is inferior to no stirring [17]. Constantly stirred samples give improved replicate precision, which increases confidence in data and lowers statistical detection limits. Samples containing significant suspended solids should be treated with caution. Municipal sewage sludges, for example, may contain 1-3% solids by weight, and when undiluted can be problematic for agitation mechanisms. Because the headspace configuration is recommended for these more complex matrices, a magnetic stirring bar provides a simple and effective stirring source. Cylindrical stirring bars should be used with care, because they tend to orient themselves to spin ineffectively around their longitudinal axis if they are not centered over the magnet or if the applied magnetic force is increased too quickly.

As an alternative, X-shaped stirbars are a viable substitute. In opaque samples such as sludges or wet sediments, a visible vortex in the sample vial is the best indicator that the stirring mechanism is functioning effectively.

Certainly no restrictions prevent adapting SPME methods outside of traditional norms to improve performance. A novel experimental technique recently demonstrated enhancement of the detection of EPA method 624 volatile compounds by combining classical static headspace analysis with headspace SPME [18]. This was accomplished by incorporating a standard gastight syringe into the fiber/syringe assembly. The result was that 110 μl of headspace gas were collected in the syringe as the fiber was retracted at the end of headspace sampling. Analyte was thus provided to a GC column from the fiber and the collected gas simultaneously during fiber re-extension inside the injector. Whereas volumetric headspace gas sampling combined with direct injection is more sensitive than headspace SPME for the early eluters, the converse is observed for the less volatile components. The two methods complement each other, maximizing sensitivity over the volatile range, when combined. The technique is not amenable for samples in small vials, however, and being early in its development, has not been automated on commercial instrumentation.

Alkyl Substituted Benzenes (BTEX)

Benzene, toluene, ethylbenzene and the xylenes (collectively referred to as BTEX) are a subset of commonly regulated volatile organic compounds, yet they compose a short list that is one of the most frequently requested in environmental monitoring. As by-products of petrochemical refining, coal burning operations, and leaks from underground storage tanks, they are routinely demanded as a unique set of parameters. It follows that a significant amount of work has been devoted to optimization of SPME extraction of these compounds. The same principals apply to BTEX analysis as to the larger set of volatiles; the preferred stationary phase is poly(dimethylsiloxane) due to its favorable polarity. A 100 μm film thickness is recommended for optimizing sensitivity; however, care must be taken with respect to the origin and type of sample. Environmental samples often contain such high analyte concentrations that a thinner film, such 56 μm, is preferable to minimize absorption of unwanted compounds. Some variability in response factors occurs within the group, in particular with benzene because its polar nature makes it less responsive to PDMS than the other BTEX components. This character is reflected by benzene's lower $\log K_{fs}$ value for a direct extraction from water, presented in Table 5.6. Consequently, salting out as a technique to increase sensitivity – in either direct immersion or headspace configurations - is the most important for benzene. Figure 5.3 demonstrates the relative responses of the BTEX components with and without salting out in the headspace extraction mode.

Table 5.6 Comparison of Distribution Constants, Extrapolated Limits of Detection (LOD), Extrapolated Limits of Quantitation (LOQ), Method Detection Limit (MDL), and Precision for BTEX Analytes in Water.

	$\log_{10}K$	LOD (ppt)	LOQ (ppt)	MDL (MISA) (ppt)	MDL (EPA) (ppt)	Precision (%)[a]	Precision (%)[b]
Benzene	2.30	15	50	500	30	7.3	5.2
Toluene	2.88	5	15	500	80	6.7	3.2
Ethylbenzene	3.33	2	7	600	60	7.2	3.6
m/p-Xylene	3.31	1	4	1100	60[c]	6.5	3.2
					30[d]		
o-Xylene	3.26	1.5	5	500	60	5.5	2.7

[a] 50 pg/mL   [c] m-Xylene
[b] 15 ng/mL   [d] p-Xylene
Source: Ref. 52.

The simplest GC detector for BTEX is the FID; it will provide adequate sensitivity for applications that do not require ultra trace levels of detection and are not exceedingly complex in terms of the matrix. Indeed, in moderate to heavy petrochemical matrices, BTEX components elute in the relatively quiescent early stage of the chromatogram. Statistically determined detection limits using a standard U.S. EPA method [19] have been established at less than 1 µg/L for each BTEX constituent by using headspace SPME with a standard FID, a 100 µm PDMS fiber, and salting out with sodium chloride [20]. Statistical MDLs are virtually always higher than those based simply upon signal-to-noise ratios; the BTEX detection limits based on a signal-to-noise ratio of 3:1 in this case would have been at least two orders of magnitude lower. The same study demonstrated a linear dynamic range of between 40-2,000 µg/L for all compounds and an excellent correlation with Purge and Trap/MS data derived from the same water samples. The wide dynamic range is a significant advantage for situations in which high- and low-concentrations occur in a single batch of samples. Precision in the middle concentration range of sensitivity is typically less than 10% RSD and can be expected to be below 5% under optimum conditions.

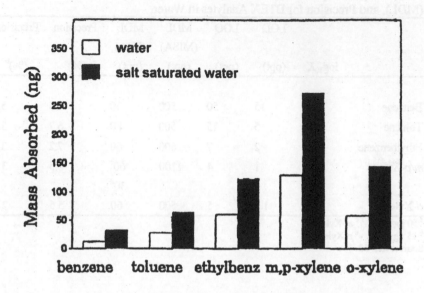

Figure 5.3 Salting out of aqueous samples, room temperature, 2 minute sampling; 1 ppm BTEX in water (white bar) and saturated aqueous salt solution (black bar). (From Ref. 16.)

Ion trap mass spectrometry delivers perhaps the best combination of high sensitivity and the ability to detect a wide range of contaminants, both in benchmark analytical methods and in SPME applications. Certainly, its capacity for low-level detection fully applies to BTEX components. Table 5.6 summarizes experimental limits of detection (LOD) and limits of quantitation (LOQ) for BTEX, comparing them to Canadian (Ontario) and EPA guidelines. Detection thresholds in water for Method 624 and 524.2 compounds at part-per-trillion concentrations have been achieved using signal-to-noise criteria. Many chemists use the convention of defining LOD at a 3:1 signal-to-noise (S/N) ratio and LOQ similarly at an S/N of 10:1. Where statistically derived values are specified by protocol, there is a requirement to perform the entire analytical procedure using a minimum number of replicates (usually 7) at identical concentrations above, but near, the estimated detection limit. The standard deviation of replicates with the student's *t*-value can then be used to calculate the estimated detection limit [19].

Gas chromatographic conditions need not be changed from those used for static headspace analysis or Purge and Trap. A minimum injector temperature of 170°C, 25°C above the boiling point of *o*-xylene, is advisable to assure efficient desorption. Cryogenic focusing of the injection band by means of introducing CO_2 or liquid nitrogen to the GC column oven during fiber desorption is an option, but cryogenics are not critical to achieving adequate separation and detection of the

BTEX components. Column diameter, length, phase, and film thickness remain the key variables at the disposal of the chromatographer for optimization of this straightforward analysis. Figure 5.4 depicts resolution of BTEX desorbed from a 100 μm PDMS fiber in 8 minutes without sub-ambient cooling. Separation of *m*- and *p*-xylene was attained in spite of the fact that generally these compounds are permitted to be reported as a coeluting pair.

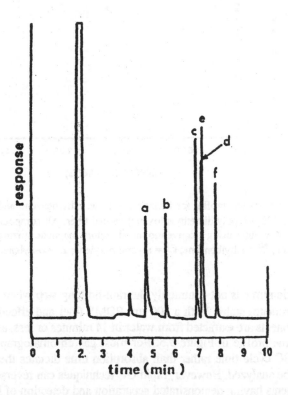

Figure 5.4 Recovery of BTEX components from water sample spiked at 170 ng/mL by SPME-CLOT-GC-FID showing resolution of the *m*- and *p*-xylene isomers. The analytes are: a. benzene, b. toluene, c. ethylbenzene, d. *p*-xylene, e. *m*-xylene, f. *o*-xylene. Unlabeled peaks to the left of benzene are methanol from the BTEX solutions used to spike the water and an unknown. Ten minutes total desorption and GC run time. (From Ref. 55.)

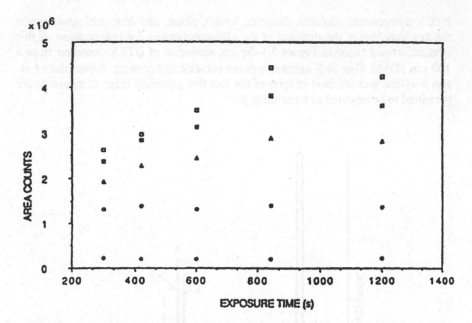

Figure 5.5 Absorption profile for benzene, toluene, ethylbenzene and xylene isomers using 1 cm of a 100 μm poly(dimethylsiloxane)-coated fiber. All compounds have reached equilibrium in 14 minutes using a conventional laboratory magnetic stirring plate. ● = Benzene; ○ = toluene; ■ = ethylbenzene; □ = *m*- and *p*-xylene; ▲ = *o*-xylene. (From Ref. 6.)

Extraction time is not commonly the rate-limiting step when using conventional gas chromatography. With a 100 μm PDMS fiber and effective stirring, all BTEX components are extracted from water in 14 minutes or less, as shown by the absorption time profile in Figure 5.5. For most gas chromatographic analyses of BTEX, the GC cycle time rather than absorption time dictates the rate at which samples can be analyzed. However, rapid GC techniques can reverse that situation with experiments having demonstrated separation and detection of BTEX components less than 20 seconds after desorption of the fiber [21]. Commercial instrumentation is accessible that will provide equivalent results.

Volatile Organics in Solid Samples

The analysis of solid samples for volatiles presents challenges regardless of the approach taken, the most critical factor is often the nature of the sample itself. Solids range from relatively homogeneous materials such as sands, clays, and lake sediments to non-uniform sludges, coarse soil samples, and industrial or oily wastes. Variables that come into play, in addition to homogeneity, include particle size and the characteristic sorption of analytes to solid particles. Additional impurities (i.e. fine stones in soil samples), if overlooked, can bias analytical results.

Because the principles of any sample preparation technique are the same (that is, they rely on partitioning between sample and an extraction medium) the character of the sample does, in practice, result in wide variations in analyte recoveries. Mention was made earlier of U.S. EPA Method 5030, which specifies two quite different pre-extraction procedures, chosen based only on the *expected* concentration of analyte in the matrix. The slurry consisting of the solid mixed with purified water is sparged in a Purge and Trap system. The liquid-solid extraction produces a methanol extract to be spiked into clean water for Purge and Trap treatment. The former method is used at analyte concentrations below 1 μg/g, the latter for all others. In fact, the two methods will frequently generate significantly different results when applied to fractions of the same solid sample, notwithstanding intrinsic variations in sub-sampling. The slurry technique assumes efficient mixing of the sample and effective transfer of analyte from the solid to the aqueous phase. In considering a relatively non-polar matrix, such as a petroleum contaminated soil for BTEX analysis, methanol extraction will often produce higher results than an aqueous slurry, particularly since BTEX components are highly soluble in methanol. With some clays, methanol facilitates disintegration of the matrix into fine particles, rendering more sample surface area available to the extracting solvent than slurrying does to a purge gas. In any case, it is necessary to keep in mind the limitations of dealing with solids when using any of the methods discussed.

In spite of apparent limitations, there is some choice in approaching the analysis of volatiles in solids, although admittedly SPME is still relatively early in its development for this application. Headspace SPME, for example, can easily substitute for Purge and Trap with either of the Method 5030 sample preparation options. When considering the methanol extraction technique, small amounts of the solvent (i.e. 50-100 μl) spiked in 5 mL of water to prepare a Purge and Trap sample will not adversely affect the aqueous/gas partitioning. If larger volumes of solvent are necessary, adding the same volume of solvent to calibration mixtures should compensate for any effects on partitioning. A simple preliminary test using replicate calibration mixtures with varying amounts of methanol will define levels below which analyte recoveries are not affected by the solvent. In headspace analysis of spiked water for BTEX, up to 5% methanol by volume in a 25 mL sample did not noticeably reduce recoveries [22].

Where slurrying is employed with headspace SPME, a magnetic stirrer might appear to achieve complete homogenization of sample, but the chemical characteristics of the solid sample (i.e. strong analyte/solid interactions) may hamper efficient transfer of analyte from liquid phase to gas phase and hence to the fiber. In the analysis of any solids, the addition of internal standards to gauge compound recoveries can be particularly useful. Deuterated analogs of analytes are the compounds of choice if mass spectrometric detection is employed. Alternately, non-interfering compounds with similar K_{fs} values may be used in sample spiking solutions to estimate recoveries of target analytes when using detectors such as FID or PID. The question as to whether internal standards added prior to sample preparation accurately represent the recoveries of analytes that are intimately ensconced in the matrix applies, regardless of whether SPME or classical methods are utilized. The value of efficient mixing, therefore, cannot be overemphasized.

Headspace above solids has been sampled directly, with modification by the addition of water to the sample vial to give 10-50% moisture in the sample [16]. Notwithstanding that the increased moisture can enhance sensitivity, it is necessary to closely control the moisture content of the solid in order to achieve reproducible results, as indicated in Figure 5.6. Caution is advised, since moisture content is a parameter that is usually fixed by the history and condition of the sample itself. There are further challenges in converting quantity of analyte measured to concentrations that are equivalent and comparable to those produced by established analytical methods. Without mixing, physical breakdown of the sample matrix is not achieved, and analytes trapped in the interior will not be released to the gas phase. Replicate sub-sampling and the ability to reproduce the same exposed sample surface area in each sampling vial are therefore key factors in attaining single digit precision. As research continues on method optimization, direct analysis of solids can perhaps best be used as a tool for screening contaminated solids for volatiles.

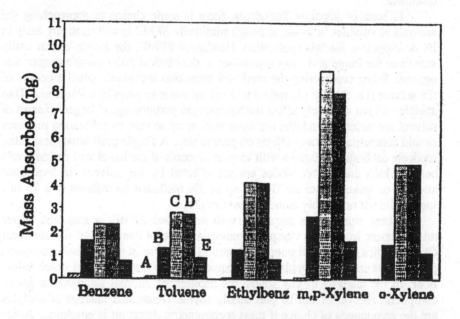

Figure 5.6 Improving the sensitivity of headspace SPME sampling by heating and matrix modification. 1ppm BTEX in clay, 2 minute sampling. A. room temperature; B. 50°C; C. 10% water and 50°C; D. 30% water and 50°C; E. 50% water and 50°C. (From Ref. 16.)

Volatile Organics in Air

Given that headspace SPME can be described as a form of air sampling, the application of a coated fiber to whole air samples involves a relatively simple modification; that is, changing the sampling chamber from a partially liquid-filled vial to a dedicated gas sampling vessel. The obvious advantages of applying SPME to air sampling are its sensitivity and low cost, especially when considering a simple arrangement in which the syringe-mounted fiber is passed through a septum and into a filled air sampling container for extraction.

In spite of the apparent simplicity of sampling air, obtaining accurate quantitative results necessitates the consideration of sample temperature and humidity which, unlike in the laboratory, cannot be controlled in the field where sampling occurs. Both parameters may have significant effects on the outcome of the analysis unless method calibration is performed under similar conditions. Poly(dimethylsiloxane) has been the most often used stationary phase in early air sampling development, though it has recently been combined with a carbon adsorbent in a hybrid fiber [23]. The 100 μm PDMS fiber is suitable for working with part-per-billion (ppb) and sub-ppb concentrations of non-polar volatiles typically measured in ambient air. Gas chromatographic column selection and temperature programming can be maintained the same as for the analysis of these compounds in water. Due to desorption being fundamentally unchanged from the water-sampling situation, cryofocusing can be avoided using the same techniques described earlier.

In comparison to current methods, the use of SPME for air sampling requires much less capital equipment. One of the benchmark techniques is EPA Method TO-1 [4] employing cylindrical sampling tubes packed with granular sorbents that have affinities for volatiles. Once the tubes have measured volumes of sample air drawn through them, the retained analytes are thermally desorbed or solvent eluted from the cartridges before analysis by GC. Sensitivity of the method is high, however, the associated hardware is expensive and can be complex to troubleshoot. In addition, excessive humidity in the sampling train affects recoveries of some volatiles and, for each sample cartridge, only one attempt at analysis is possible. To be useful, a cartridge must deliver a quantity of analyte within the handling range of the analytical instrumentation. As a result, the field sampling technician must either acquire multiple samples at differing volumes, or have a prior estimate of analyte concentrations in order to adjust the volume sampled.

A second accepted technique for the analysis of toxic organics in ambient air samples is U.S. EPA Method TO-14 [4], which utilizes stainless steel collection canisters having deactivated internal surfaces to minimize losses of analytes to sorption. A variation of the technique involves substitution of canisters by sampling bags constructed of custom materials to reduce adsorption and analyte diffusion through their walls [24]. In Method TO-14, a volume of air is removed quantitatively from a steel canister and introduced to a gas chromatograph. A cryofocusing stage in-line condenses and thus concentrates the analytes while allowing the sample air to pass through to a vent. A schematic of a TO-14 analytical system is given in Figure 5.7. The sample path is relatively complex; therefore, the operator must be on guard against potential analyte losses to wetted surfaces, including metal tubing and plumbing fixtures such as connectors and valves. The detection limits depend upon the mode of detection used (MS, PID, ECD, FID) and vary inversely as the volume of air sampled from the canister. Detection limits for selected compounds by TO-14 with GC and a photoionization detector have been estimated and compared to some SPME literature values in Table 5.7. The accompanying data indicates that equilibration times for most compounds are rapid in comparison to sampling from liquid, chiefly due to diffusion coefficients in gas being approximately four orders of magnitude greater than in water. A 10 minute sampling time using a 100 μm PDMS fiber is adequate for many compounds that have been investigated so far, including n-undecane, which has a fiber/air partition coefficient of 25,000 [25]. Since GC analyses still typically require a cycle time of greater than 10 minutes, sampling is not rate determining.

Table 5.7 SPME Equilibration Times and Comparison of MDLs with Method TO-14 for Selected Chlorinated Volatile Compounds in Air.

| | | MDL (ppb v/v) | | |
Compound	Equilibration Time (minutes)	TO-14 based on 0.5 L Sample (PID)	Experimental by SPME[a] (ECD)	Experimental by SPME[b]
Chloroform	1	0.8	0.9	2
1,1,1-Trichloroethane	3	0.7	0.5	2
Carbon tetrachloride	2	0.6	0.1	1.7
Tetrachloroethylene	5	0.02	0.01	0.05

[a] Source: Ref. 51
[b] Source: Ref. 10

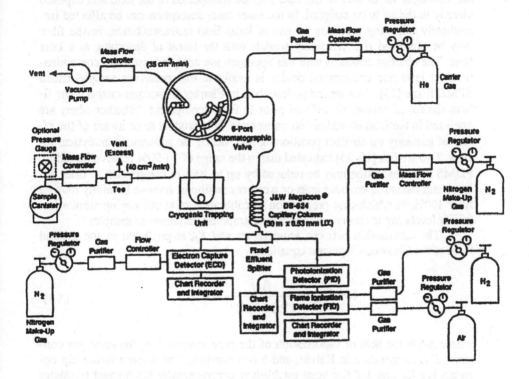

Figure 5.7 System configuration associated with the GC-FID-ECD-PID analytical system with the 6-port valve in the sample desorption mode. (From Ref. 16.)

Calibration and quantitation of SPME air samples merit particular attention. Air samples can either be collected in volumetric sampling vessels and returned to the laboratory for SPME, or the fiber may be transported to the field and exposed directly to the air to be sampled. In the latter case, desorption can be affected immediately following sampling by use of local field instrumentation, or the fiber may be preserved (i.e. efficiently sealed) with the intent of desorbing at a later time. The obvious concerns with this approach are analyte losses and contamination. At least one commercial device is available that preserves sample-loaded SPME fibers [23]. This device is described in Chapter 3. Storage methods for fibers should, of course, be verified prior to field deployment. Whether fibers are analyzed in the field or sealed and transported, it is prudent to be aware of the effect of humidity on air/fiber partitioning and, hence, on subsequent analytical results. Two studies [10, 25] indicated that in the range of 75-90% relative humidity, PDMS fiber absorption may be reduced by up to 10% in comparison to drier air. In certain industrial environments or weather conditions, relative humidity may be fully 100%, in which case preparation of calibration mixtures having similar humidity levels can in theory compensate for response variations in samples.

The relationship between temperature and K_{fh} is predicted in the general form of the Clausius-Clapeyron equation:

$$\log K_{fh} = \frac{\Delta H}{(2.303)RT} + b \tag{5.3}$$

where ΔH is the heat of vaporization of the pure analyte, R is the ideal gas constant, T is temperature in Kelvin, and b is a constant. The linear relationship between $\log K_{fh}$ and $1/T$ has been established experimentally [25,26] and translates into a 20 to 45% decrease in K_{fh} for each 10 °C increase in temperature. In principle, it is possible to calculate airborne concentrations of analytes at varied temperatures without repeated calibration of the sampling device if K_{fh} for a fiber is known as a function of temperature. As a further step, mathematical models relating K_{fh} to well-known and established chromatographic retention index data [26, 27] can eliminate the need even to determine K_{fh} experimentally; it can be calculated from existing data in the literature. This research has indicated that theoretical calculations are more in agreement with some classes of compounds than others, and that there can be some uncertainty in calculated results due to unexpected dependencies of K_{fh} on concentration [26]. In spite of their limitations, these methods show promise for application in specific situations. As they are further validated, it is expected that they will demonstrate the ability to consistently meet the requirements for accuracy and precision set by end users.

Although a rigorous calibration process can be labor intensive and expensive, calibration of the analytical system by exposing a fiber to standard gas mixtures of analytes under controlled laboratory conditions provides confidence in quantitative results which is second to none. Nevertheless, this approach must be used carefully, because the possibility exists for depletion of analytes from gas mixtures as replicates are extracted. If the volume of the gas sampling vessel is ≥ 1.0 liter or if $K_{fh} < 1,000$, several replicates will not significantly change the analyte concentration. Replicates taken under such conditions should provide relative standard deviations of 10% or less. Moreover, a controlled calibration does not preclude a field sample taken at a different temperature or humidity than that of the calibration mixture. Adjustments in methods or initial investigative work may therefore be necessary. The addition of internal standards to whole air samples arriving from the field provides extra confidence in the analytical process and should be applied where possible.

Semivolatile Compounds

Applications of SPME to semivolatile compounds are potentially wide ranging. To date, the matrices investigated for SPME applications have most often been aqueous; however, quantitative analysis of solids has been demonstrated. Fibers may be used to sample air or headspace for many of these compounds whose high gas/fiber partition coefficients compensate to varying degrees for their lower vapor pressures. The general trend of higher K_{fs} values in the semivolatile compounds enhances sensitivity in aqueous sampling, but there are constraints, as well. Equilibration times are increased, and complete desorption of analytes from fibers can be difficult to achieve for selected analytes. As always, establishing the performance of the fiber over periods of continued use is important to ensure accuracy and consistency of data.

A significant volume of SPME developmental work has been completed for organic semivolatiles, often referred to as *extractables*. The task of testing and validating SPME methods for the extensive lists of analytes that are currently monitored is an ongoing one. In addition to work with organics, progress is being made in inorganic applications such as metals analysis, which comprises a significant component of environmental laboratory testing. The subject of semivolatiles is broad so only a few of the most commonly requested analyte classes will be discussed here. By no means are the techniques limited by boundaries of current investigations; novel applications continue to emerge as needs arise.

Pesticides

Here, the term *pesticides* will be broadly defined as the class of compounds that includes insecticides, herbicides, fungicides, rodenticides, and others. In excess of a thousand such compounds have been manufactured and marketed commercially, rendering it impossible to measure all of them routinely. In fact, monitoring any more than a fraction of the total is impractical; therefore, I have focussed on pesticides deemed to be in widest use and of the greatest threat to human health. North American regulations contain dozens of official methods encompassing pesticides in water, wastewater, and solids, specifying analysis by either liquid or gas chromatography. Several of the methods pertain to a single compound, posing a further challenge to economical monitoring.

For practical purposes, agencies in the United States and elsewhere have established several categories of pesticides, usually based on chemical composition or functional groups. The distinctions conveniently allow for the analysis of like compounds on dedicated GC detectors. Nitrogen/phosphorus-containing pesticides and organochlorine pesticides are two of the most common categories, though by no means do they represent all pesticide chemistries. Both of these classes of pesticides have U.S. EPA drinking water methods dedicated to them, each specifying liquid-liquid extraction (LLE) and gas chromatography. Method 507 addresses nitrogen-phosphorus compounds and Method 508 applies to organochlorines [1]. In recent years, solid phase extraction (SPE) has emerged as a valid alternative to reduce solvent usage and labor.

Solid Phase Microextraction strives to simplify the analysis still further, having been found to be extremely well suited to the analysis of pesticides. The relatively high octanol-water coefficients of pesticides correspond to high values of K_{fs}. Because most compounds have K_{fs} values in excess of 1000, the possibility of achieving below part-per-trillion (ppt) detection is achievable in water samples for certain compounds using either electron capture detection or MS detectors in selected ion mode. There is no sample clean up in a direct immersion extraction; therefore, MS detection may, in some circumstances, be required for unambiguous analyte identification, particularly where an FID or NPD (nitrogen-phosphorus detector) is the primary detection device.

The preferred fiber coating for the largely non-polar organochlorine pesticides is 100 µm PDMS, whereas the more polar nitrogen/phosphorus pesticides are suited to a polyacrylate stationary phase, typically 85µm in thickness. Enhancements to sensitivity can be realized by adjusting ionic strength, pH, or both. Generally the organochlorines are not significantly affected by lowering pH; however, addition of NaCl to samples can improve analyte responses. Predictably, both adding salt and lowering sample pH enhances acid herbicide extractions, but the combination can reduce the amount of nitroaniline herbicides extracted [28]. Adjustment of pH and ionic strength are not always necessary to meet the required performance criteria, and since individual components among the suite of pesticides in a given analyte list have varied optimum conditions for extraction, such tools are at the disposal and judgment of the analyst.

The higher trend in fiber/sample partition ratios does not unduly lengthen equilibration times, particularly when compared to the extraction and sample work-up required by LLE and, to a lesser extent, SPE. Experimentally determined equilibration times and K_{fs} values for selected organochlorine pesticides and organonitrogen herbicides are summarized in Tables 5.8 and 5.9. Seldom is a 180 minute extraction time feasible, and for compounds having exceedingly high values of K_{fs}, shorter extraction times (i.e. 20-30 minutes) will often deliver adequate sensitivity. Similar equilibration time data has been collected for selected organophosphorus compounds [29], demonstrating complete extractions in 15-45 minutes.

Researchers have reported detection limits in many cases that meet or exceed U.S. EPA method guidelines for pesticides/herbicides in water. In most cases, either the 100 μm PDMS or the 85 μm polyacrylate fiber will permit regulatory detection limits to be surpassed in drinking water, provided that the mode of detection is sufficiently sensitive. Table 5.10 compares experimental detection limits of nitrogen/phosphorus pesticides with those cited in EPA Method 507. Comparison of data detection limit data is again limited by the fact that the EPA method for estimating detection limits would ordinarily produce higher values than the experimental signal-to-noise values under equal instrument conditions. In spite of the differences, the SPME data are an order lower than those in Method 507, indicating that method sensitivity is quite respectable.

Linearity of the method is essentially detector dependent, as is the case for traditional sample preparation methods. Linear ranges of 0.001-100 ng/mL have been reported for organochlorines using MS detection, whereas the same compounds exhibit linearity only between 0.001 and 1 ng/mL when an ECD is used [30]. For nitrogen-containing herbicides, a method employing a polyacrylate fiber has demonstrated linearity between 0.1-1000 ng/mL using any of NPD, FID and MS detection systems [28]. Precision is comparable to that given in Methods 507/508, which indicate RSDs below 20% for virtually all parameters. Precision lower than 20% has been reported for organochlorines [30] and less than 10% for nitrogen/phosphorus compounds [31].

Because direct immersion of the fiber is necessary for exposure to the majority of pesticides and herbicides in water, it is important to establish the effects of competing species on absorption, as well as the robustness of a fiber exposed to dirty matrices. Part-per-million (μg/mL) levels of extraneous organics should not adversely affect the partitioning of pesticides into 85μm or 100μm stationary phases. When exposed to certain matrices, particularly wastewaters, fibers may visibly discolor without apparent changes in performance; in other situations they may lose some sorptive capacity [32].

Gas chromatographic conditions and columns for SPME/GC systems vary and are best determined by the characteristic target parameters. Figures 5.8 through 5.10, provide examples of extraction and chromatographic conditions for subsets of pesticides/herbicides. Although most SPME work with pesticides has been combined with GC analysis, some carbamates and urea pesticides have effectively been extracted and analyzed by HPLC [33]. The analysis requires an SPME/HPLC hardware interface, which will be described in more detail for PAH applications.

Table 5.8 Properties for Target Organochlorine Pesticides.

Pesticide	Equilibration Time (minutes)	K_{fs} Values
Hexachlorocyclohexanes		
a-BHC	15	1800
β-BHC	15	900
δ-BHC	15	600
Lindane (γ-isomer)	15	13000
Diphenyl aliphatics		
Methoxychlor	180	18000
p,p'-DDD	180	21000
p,p'-DDE	90	10000
p,p'-DDT	180	23000
Cyclodienes		
Aldrin	180	10000
Dieldrin	120	25000
Endosulfan I	45	25000
Endosulfan II	45	10000
Endosulfan sulfate	45	400
Endrin	120	21000
Endrin aldehyde	120	1400
Endrin ketone	45	2300
Heptachlor	180	18000
Heptachlor epoxide	180	35000

Source: Ref. 30.

Table 5.9 Selected Nitrogen Containing Herbicides.

Herbicide Class	Name	Equilibration Time (min)	K_{fs} Value
Triazines	Atrazine	90	2000[a]
	Hexazinone	10	300[a]
	Propazine	90-120	3000[a]
	Metribuzin	50-90	200[a]
	Simazine	10	300[a]
Nitroanilines	Benfluralin	50-90	7000[b]
	Isopropalin	50	5000[b]
	Pendimethalin	50	20000[b]
	Profluralin	30	7000[b]
	Trifluralin	50	8000[b]
Substituted Uracils	Bromacil	30	400[a]
	Terbacil	50-90	200[a]
Thiocarbamates	EPTC	90	4000[a]
	Molinate	50-90	2000[a]
	Cycloate	90	7000[a]
	Butylate	90	3000[a]
	Pebulate	90	4000[a]
	Vernolate	90	10000[a]
Others	Metolachlor	50	4000[a]
	Oxyfluorofen	30	3000[b]
	Oxadiazon	50	20000[b]
	Propachlor	50	1000[a]

[a,b] K_{fs} values determined under optimum conditions; overnight extraction, 4 M NaCl ([a]) or unsalted ([b])
Source: Ref. 28.

Table 5.10 Detection Limits of Selected Nitrogen/Phosphorus Pesticides.
Detection Limit (ng/L)

Pesticide	SPME Experimental LOD[1]	EPA Method 507[2]
Alochlor	22	380
Ametryn	9	200
Atrazine	10	130
Diazinon	3	250
Metolachlor	13	750
Metribuzin	110	150
Prometon	8	300
Prometryn	9	190
Simetryn	9	250
Terbutryn	15	250

[1] Determined by MS detection (SIM) mode and based signal-to-noise ratio of 3:1. Source: Ref. 31.
[2] Determined by ECD using EPA method (19) or signal-to-noise ratio of 5, whichever is higher.

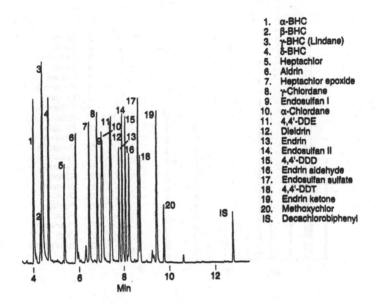

1. α-BHC
2. β-BHC
3. γ-BHC (Lindane)
4. δ-BHC
5. Heptachlor
6. Aldrin
7. Heptachlor epoxide
8. γ-Chlordane
9. Endosulfan I
10. α-Chlordane
11. 4,4'-DDE
12. Dieldrin
13. Endrin
14. Endosulfan II
15. 4,4'-DDD
16. Endrin aldehyde
17. Endosulfan sulfate
18. 4,4'-DDT
19. Endrin ketone
20. Methoxychlor
IS. Decachlorobiphenyl

Figure 5.8 Separation of chlorinated pesticides after SPME. 200 ppt each in 2 mL water. Direct extraction for 15 min. with rapid stirring using a poly(dimethylsiloxane) fiber, 100 μm film. Separation on SPB-5 column, 15 m x 0.20 mm, I.D., 0.20 μm film. Column oven program is 120°C for 1 min, to 180°C at 30°/min, then to 290°C at 10°/min. Helium carrier gas at 37 cm/sec (set at 120°C). Splitless injection at 260°C. ECD at 300°C. (Courtesy of Supelco, Bellefonte, PA.)

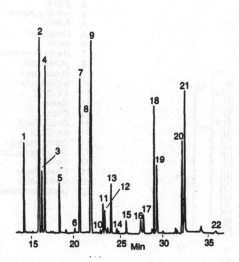

1. Eptam
2. Sutan
3. Vernam
4. Tillam
5. Ordram
6. Propachlor
7. Ro-Neet
8. Trifluralin
9. Balan
10. Simazine
11. Atrazine
12. Propazine
13. Tolban
14. Terbacil
15. Sencor
16. Bromacil
17. Dual
18. Paarlan
19. Prowl
20. Oxadiazon
21. Goal
22. Hexazinone

Figure 5.9 Nitrogen-containing herbicides extracted from water. Water sample contains 100 ng/mL each analyte + 1 g/mL NaCl, pH 2 in a 4.6 mL vial. Polyacrylate, 85 μm film. Direct extraction at ambient temperature for 50 min., constant stirring. Desorption for 5 min at 230°C into a splitless injector. Column is PTE-5 , 30 m x 0.25 mm, I.D., 0.25 μm film. Column oven program is 40°C for 5 min. to 100°C at 30°/min, then 5°/min to 275°C. Helium carrier at 40 cm/sec (set at 40°C). Ion Trap MS detection at 250°C, mass scan range: 45-400 m/z at 0.6 sec/scan. (Courtesy of Supelco, Bellefonte, PA.)

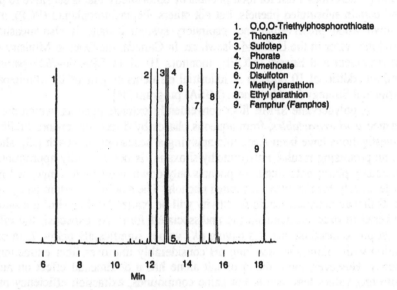

1. O,O,O-Triethylphosphorothioate
2. Thionazin
3. Sulfotep
4. Phorate
5. Dimethoate
6. Disulfoton
7. Methyl parathion
8. Ethyl parathion
9. Famphur (Famphos)

Figure 5.10 Separation of organophosphorus pesticides after SPME. Sample is 50 ppb of each analyte in 1.5 mL saturated salt water, pH 7.2. 20 min direct extraction with rapid stirring. Poly(dimethylsiloxane) fiber, 100 μm film. Column is PTE-5, 30 m x 0.25 mm I.D., 0.25 μm film. Column oven temperature is 60°C for 1 min, 12°/min to 300°C, hold 5 min. Helium carrier at 30 cm/sec. Splitless injection (0.75 mm liner) at 270°C. MS detection: m/z is 45-400, 0.6 sec/scan. (Courtesy of Supelco, Bellefonte, PA.)

Phenols

Phenols are monitored largely in industrial and municipal wastewaters, and have been typically approached using either colorimetric means or using liquid-liquid extraction/GC. The U.S. EPA methods 604/625 [2] provide LLE methods for separation and measurement of 11 phenols in wastewater. Standard Method 5530 [5] describes a test for total phenols by colorimetry that is sensitive to phenol and certain substituted phenols, but not others. Pentachlorophenol (PCP), a well-known wood preservative and respiratory system irritant, is also measured in drinking water in the U.S. and elsewhere. In Canada, the Ontario Ministry of the Environment and Energy (MOEE) monitors 10 of 11 EPA 604/625 parameters and an additional 10 phenols in industrial effluents as part of its Municipal and Industrial Strategy for Abatement (MISA) program [34].

A polyacrylate SPME fiber conveniently extracts phenols, which are often termed *acid extractables*, from aqueous phases by direct immersion. Preliminary investigations have been made into applying a headspace approach [35], showing some promising results. Poly(dimethylsiloxane) is not generally appropriate as a stationary phase, detecting most phenols only down to the mg/L range, and having little affinity for the nitro-substituted phenols. The acidity of certain analytes suggests that an enhancement in sensitivity will be realized if the pH of the sample is lowered in order to maintain the undissociated form. As expected, the effect is most pronounced on analytes having pK_{fs} values significantly below 7. In combination with salting out, lowering pH considerably improves fiber extraction efficiency. However, the addition of salt alone has a detrimental effect on analytes with pK_{fs} values less than 8. For some compounds, extraction efficiency may be reduced by up to an order of magnitude.

Sensitivity required by regulations can be met or exceeded when capable detectors are used, such as GC/MS in the selected ion mode. Precision has been demonstrated to be impressive; Table 5.11 indicates an average RSD of less than 10% for selected phenols. The data further indicate sub-part-per-billion detection limits, defined by signal-to-noise criteria, for virtually all compounds examined. Caution should be taken with dinitrophenols, and particularly 2,4-dinitrophenol, which in its normal free acid form can experience chromatographic losses even during simple solvent injections of reference mixtures. Some variability in these components is to be expected, and it relies heavily on both the matrix and the cleanliness of the GC injector/column path.

Table 5.11 Detection Limits and Precision for Selected Phenols.

Compound	SPME[a,b]	MISA	Method 604 GC-FID	Method 625 GC/MS	SPME Precision (% RSD)[b]
phenol	0.80	2.4	0.14	1.5	4.2
2-chlorophenol	0.24	3.7	0.31	3.3	4.2
o-cresol	0.09	3.7	-	-	-
m-cresol	0.11	3.4	-	-	-
p-cresol	0.11	3.5	-	-	-
2,4-dimethylphenol	0.02	7.3	0.32	2.7	4.8
2,4-dichlorophenol	0.02	1.7	0.39	2.7	4.9
2,6-dichlorophenol	0.01	2.0	-	-	-
4-chloro-3-methylphenol	0.01	1.5	0.36	3.0	4.0
2,3,5-trichlorophenol	0.02	1.3	-	-	-
2,4,6-trichlorophenol	0.08	1.3	0.64	2.7	4.5
2,4,5-trichlorophenol	0.07	1.3	-	-	-
2,3,4-trichlorophenol	0.04	0.6	-	-	-
2,4-dinitrophenol	1.6	42.0	13.0	42.0	8.9
4-nitrophenol	0.75	1.4	2.8	2.4	9.3
2,3,5,6-tetrachlorophenol	0.01	1.6	-	-	-
2,3,4,5-tetrachlorophenol	0.01	0.4	-	-	-
2,3,4,6-tetrachlorophenol	0.01	2.8	-	-	-
2-methyl-4,6-dinitrophenol	0.44	24.0	16.0	24.0	5.6
pentachlorophenol	0.11	1.3	7.4	3.6	12

[a] GC/MS detection.
[b] Source: Ref. 35.

Initial conditioning of the polyacrylate fiber is recommended prior to analysis to remove intrinsic contaminants. This may be done at 300°C under a steady flow of helium by leaving the fiber exposed in GC injector port for 10-20 minutes, making sure to bake the column until a clean detector baseline is observed. Carryover can be observed with some phenols; however, it is likely to be most significant for pentachlorophenol. Preliminary tests are advisable and may lead to increased injector temperatures and longer desorption times. Injection temperatures of 250-280°C and desorption times of 5-7 minutes should be sufficient to obtain acceptable fiber blanks following a calibration or analysis.

Linearity over two orders of magnitude for most compounds should be attainable using a FID, and a greater range for some compounds is feasible with MS detection. Temperature programs and columns can be similar to those used in conventional syringe injections. Chromatographic acquisition time is typically 25-30 minutes.

Polynuclear Aromatic Hydrocarbons

Extensive literature has been compiled on this class of environmental contaminants, both in terms of analytical methodologies and origin/transport/fate. Comprising a sub-section of the neutral/extractables class, PAHs and are ubiquitous in the environment. They have been identified as a health concern as a result of toxicological studies, and benzo(a)pyrene in particular has been identified as a human carcinogen. Current sample preparation methods use liquid-liquid or liquid-solid extraction for aqueous samples and various liquid-solid solvent extraction approaches for soils, sediments and related solids. The U.S. EPA addresses PAHs in wastewaters in Methods 610/625 [2], in drinking water in Method 525 [1], and in liquid and solid hazardous wastes by various preparation methods [3].

Methods employing SPME are easily adaptable to polynuclear aromatics in aqueous solutions. Most PAHs, and especially those compounds with greater than 4 rings, have high K_{fs} values, resulting in high sensitivity. Their non-polar character makes them ideal for extraction using a poly(dimethylsiloxane) stationary phase. The favorable partitioning ratio is indicated by pK_{fs} values generally above 3, and for many applications a thinner coating (7μm or 30 μm) is necessary not to reduce equilibration time, but rather to reduce the amount of absorbed analyte and the tendency for sample carryover. Reducing fiber exposure time still further may be necessary if concentrations are exceedingly high; a simple alternative is to merely dilute the sample. Thicker fiber coatings have value where extreme sensitivity is required and the sample carryover problem can be solved, possibly by increasing GC desorption time and injector temperature, or by utilizing solvent desorption of the fiber.

Extraction from liquid samples is typically done by direct fiber immersion. After extraction, there is a choice of applying either gas or liquid chromatography for the analytical separation and measurement. Gas chromatography of PAHs is a firmly established technique and routine methods perform the separation of 16

U.S. EPA Method 610 PAHs in 30 minutes or less. The relative inertness of the components reduces susceptibility to peak tailing, and the separation can often perform well even on an analytical column that has been inundated with dirty samples. Some carryover of the heavier PAHs is almost inevitable with the use of a 100 μm stationary phase, and the thick coating should be avoided if a thinner film will suffice. Even with a thinner film, carryover should be monitored for the less volatile compounds such as benzo(a)pyrene and benz(a)anthracene. It has been indicated that raising GC injector temperatures over 300°C can enhance desorption and reduce carryover of PAHs from PDMS fibers without fiber bleed or thermal degradation, if the stationary phase is manufactured carefully [7].

A linear dynamic range of 3-5 orders of magnitude should be observed if detected by a FID or ion trap mass spectrometry in the selected ion mode. The latter detector has been used to report measurement of selected PAHs below 10 pg/mL [7]. Such performance is comparable to EPA Method 525 for drinking water, which does not require method detection limits below 30 pg/mL for any of its target analytes. The same study determined precision of anthracene, benz(a)anthracene, and benzo(a)pyrene at 0.25 μg/L to a range from 8-11 % RSD; the RSD values for these compounds by method 525 using a solid phase C_{18} extraction cartridge are 7.7, 7.1 and 33%, respectively [1].

The use of a liquid chromatography (HPLC) system to desorb and separate PAHs is a relatively recent development, offering an alternative to GC. A modified injection system replaces the conventional injection loop and effectively solvent desorbs the analytes. Figure 5.11 illustrates the design of a commercial SPME/HPLC interface. Liquid chromatography columns for PAH separation can be readily purchased. Sample carryover seems to be overcome with this form of solvent elution [36]; however, the sensitivity realized with HPLC/UV detection is significantly less than achievable by GC/ion trap MS. Naphthalene and the three ring PAHs have been measured with reasonable facility using LC to the 10 μg/L level and 4- to 6-ring PAHs to 1-2 μg/L [24]. Continuing research with LC instrumentation should lead to both further enhancements in sensitivity, as well as the production of detailed performance data.

Applying SPME to PAHs in solids and soils is less straightforward. Avoiding time-consuming sample extractions that rely heavily on solvents has been one of the chief advantages of SPME; yet most benchmark solid sample preparations have, to this point, relied on some form of solvent extraction. One proposed procedure that requires no organic solvent is a hot water extraction, which exposes subcritical water to a sample in a closed extraction vessel [37]. The subsequent SPME step is applied to the aqueous phase after cooling and removal from the vessel. The method compared reasonably well with soxhlet extractions, having a tendency for higher relative recoveries for the more volatile PAHs. For consistent quantitation, it is necessary to spike the sample with isotopically labeled surrogates for each analyte prior to heating the extraction vessel. The surrogates account for re-sorption of the non-polar analytes by the solid matrix while the extraction vessel is cooled.

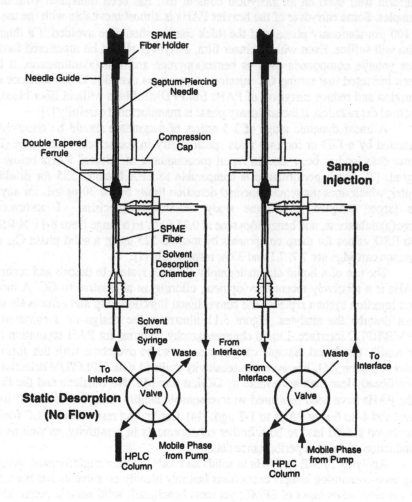

Figure 5.11 Schematic of an SPME/HPLC interface in desorption mode and injection mode. (Courtesy of Supelco, Bellefonte, PA.) Source: Ref. 54.

In terms of gas phase analysis of PAHs, relatively little has been written, yet SPME may find applications in this area. Limited success has been had with headspace analysis of PAHs from water samples in the laboratory, largely due to their lack of volatility. The U.S. EPA and others routinely monitor vapor phase PAHs in ambient air using polyurethane foam traps coupled to high volume air sampling pumps, specified in Method TO-13 [4]. Assuming the necessary sensitivity can be demonstrated, screening industrial or ambient air by SPME will prove to be valuable for circumventing unnecessary allocation and deployment of expensive air sampling equipment.

Other Applications

Other methods and applications being developed for Solid Phase Microextraction of semivolatiles are at earlier stages in their development. Certainly SPME has been applied to extractable compounds other than those mentioned above. One study [38] employing PDMS fibers investigated the range of compounds contained in EPA method 1625 [2], an isotope dilution method, which specifies over 80 target analytes representing basic, neutral, and acidic compounds. All but 16 of the compounds investigated were detected in water down to 10 μg/L, reinforcing the fact that PDMS is applicable to expansive target lists. Other fiber chemistries will no doubt prove to be effective for the more polar Method 1625 parameters.

The analysis of nonionic surfactants in water is another test amenable to fiber extraction. These materials, comprising commercial detergents, often in percent quantities, have been linked to effects on the human endocrine system and have been detected in sewage treatment plant effluents. A recently published method for determining alkylphenol ethoxylates in water using SPME/HPLC has led to commercializing a fiber coating dedicated to the extraction of surfactants [39]. The 3:1 signal-to-noise detection threshold is in the low parts-per-billion region and 3 orders of linearity can be expected with UV detection. Precision has been reported at less than 15%, comparable to most regulatory precision requirements.

An emerging niche for SPME is in the field of fuel related hydrocarbons, particularly total petroleum hydrocarbons (TPH) analysis, because the PDMS fiber is an excellent extraction medium for virtually all non-polar hydrocarbons in both water and air. In addition to BTEX and PAHs, n-alkanes up to decane have been extracted from surface waters and sewage treatment plant effluents [40] and undecane has been extracted from air [25]. Gasoline, diesel, and kerosene have all been effectively extracted from water. The challenge in applying new methods to TPH measurement, however, will be less technical than procedural. Regional and international disparities between scientists on definitions and equivalency associated with TPH methods has resulted in some confusion with respect to the significance of data being generated. As TPH issues are resolved, it is reasonable to assume that SPME will have a significant impact on the measurement of this class of compounds.

Initial investigations have been made with other analytes that require further validation to gain widespread acceptance, including fatty acids by *in situ* derivatization/SPME [41], PCBs [7], organometallic compounds by FID and atomic emission detection [42,43], and explosives. Researchers have also applied atypical analytical configurations to the determinations of familiar compounds. Volatile organics have been detected by infrared spectroscopy [44], BTEX has been measured using Raman spectroscopy [45], and capillary electrophoresis was applied to the measurement of PAHs following SPME [46]. In addition, sub-ambient cooling has been applied in situ to an SPME fiber to enhance partitioning ratios [47] toluene has been monitored in a flowing stream [48], and a process to remove BTEX from air has been evaluated using SPME [49]. While many novel approaches will not be practical for use across the environmental sector due to restricted sensitivity or other limitations, they demonstrate the adaptability of SPME to a diverse suite of analyses.

FUTURE DEVELOPMENTS

Although somewhat speculative, future efforts in the application of SPME will focus on reducing time of analysis for many tests, proofing field analytical methods, and establishing preservation effectiveness of field samples. Systems employing rapid GC coupled with SPME [50] can now be purchased and should find increased use. Commercial autosampling options may expand to provide the option of agitation of the aqueous phase for headspace samples, or complete air analysis.

Stationary phase development is continuous, and new phases becoming available will continue to address specialty analyte applications in much the same way that capillary GC columns have done. Indications are that PDMS will continue to be combined with other phases for niche applications. Extraction of polar organics from water is expected to improve. Ion exchange polymers will be further explored as investigators apply them to extract ionic species. Moreover, there is interest by some users in obtaining low-cost, disposable fibers, provided they can be manufactured uniformly. Should they come available, they will find applications in screening dirty samples and in simultaneous field sampling where multiple samples are required.

Proposals will undoubtedly be made to regulatory agencies with the intent of having SPME adopted into official methods. To that end, efforts at validating specific methods and applications will be increasingly formalized. In the interim, SPME will continue to supply alternatives to conventional methods of analysis.

REFERENCES

1. Methods for the Determination of Organic Compounds in Drinking Water, EPA/600/4-88/039, U.S. Environmental Protection Agency, Cincinnati, OH, 1988.
2. Code of Federal Regulations, Title 40, Pt. 136, Appendix A, Methods 601-1625, U.S. Government Printing Office, Washington, D.C., 1988.
3. Test Methods for Evaluating Solid Waste, SW-846, U.S. Environmental Protection Agency, Washington, D.C., 1995.
4. Compendium of Methods for the Determination of Toxic Organic Compounds in Ambient Air, EPA/600/4-89-017, U.S. Environmental Protection Agency, Atmospheric Research and Exposure Assessment Laboratory, NC, 1988.
5. Standard Methods for the Analysis of Water and Wastewater, 19th Ed. (AD Eaton, LS Clesceri, and AE Greenberg, eds.) American Public Health Association, Washington, D.C., 1995.
6. DW Potter and J Pawliszyn. J Chromatogr. 625: 247, 1992.
7. DW Potter and J Pawliszyn. Environ Sci Technol 28: 298-305, 1994.
8. Bulletin 901: Solid Phase Microextraction/Capillary GC Analysis of Drugs, Alcohols and Organic Solvents in Biological Fluids, Supelco Inc., Bellefonte, PA, 1996.
9. C L Arthur, LM Killam, S Motlagh, M Lim, DW Potter, and J Pawliszyn. Environ Sci Technol 26: 979-983, 1992.
10. M Chai and J Pawliszyn. Environ Sci Technol 29: 693-701, 1995.
11. Automation with Agitation, SPME Highights. Supelco Inc., Bellefonte, PA, Summer, 1996.
12. Z Zhang and J Pawliszyn. Anal Chem 65: 1843-1852, 1993.
13. T Nilsson, F Pelusio, L Montanarella, B Larsen, S Facchetti, and JO Madsen. J High Resol Chromatogr 18: 617-624, 1995.
14. R Shirey. J High Resol Chromatogr 18: 495-499, 1995.
15. C Arthur, K Pratt, S Motlagh, J Pawliszyn, and RP Belardi. J High Resol Chromatogr 15: 741-744, 1992.
16. Z Zhang and J Pawliszyn. J High Resol Chromatogr 16: 689-692, 1993.
17. Optimizing SPME: Parameters for Control to Ensure Consistent Results, Application Note 95, Supelco Inc., Bellefonte, PA, 1996.
18. Z Zhang and J Pawliszyn. J High Resol Chromatogr 19: 155-160, 1996.
19. Code of Federal Regulations, Title 40, Pt. 136, Appendix B, U.S. Government Printing Office, Washington, D.C., 1988.
20. B MacGillivray, J Pawliszyn, P Fowlie, and C Sagara. J Chromatogr Sci 32: 317-322, 1994.
21. T Górecki and J Pawliszyn. J. High Resol. Chromatogr. 18: 161-166, 1995.
22. B MacGillivray, M. Sc. Thesis, University of Waterloo, 1995.
23. R Shirey, V Mani, and W Betz. The Reporter, Supelco, Bellefonte, PA, Vol. 16, No. 4: 3, 1997.
24. C Woolley and R Mindrup. The Reporter, Supelco, Bellefonte, PA, Vol. 15, No. 3: 6, 1996.
25. P Martos and J Pawliszyn. Anal Chem 69: 206-215, 1997.
26. R Bartlett. Anal Chem 69: 364-372, 1997.
27. P Martos, A Saraullo, and J Pawliszyn. Anal Chem 69: 402-408, 1997.
28. AA Boyd-Boland and J Pawliszyn. J Chromatogr A 704: 163-172, 1995.
29. S Magdic, AA Boyd-Boland, and J Pawliszyn. Organohalogen Compd. 23: 47, 1995.
30. S Magdic and J Pawliszyn. J Chromatogr A 723: 111-122, 1996.

31. R Eisert and K Levsen. J Am Soc Mass Spectrom 6: 1119-1130, 1995.
32. KN Graham, LP Sarna, GRB Webster, JD Gaynor, and HYF Ng. J Chromatogr A 725: 129, 1996.
33. G Gora-Maslak and V Mani. The Reporter, Supelco, Bellefonte, PA, Vol. 16, No. 3: 5, 1997.
34. J Phyper, B Ibbotson. The Handbook of Environmental Compliance in Ontario, McGraw-Hill Ryerson, Toronto, 1991.
35. KD Buchholz and J Pawliszyn. Anal Chem 66: 160-167, 1994.
36. J Chen and J Pawliszyn. Anal Chem 67: 2530-2533, 1995.
37. KJ Hageman, L Mazeas, CB Grabanski, DJ Miller, and SB Hawthorne. Anal Chem 68: 3892-3898, 1996.
38. BN Colby and GM Simcik. Solid Phase Microextraction Preliminary Study, Pacific Analytical, Carlsbad, CA, 1993.
39. AA Boyd-Boland and JB Pawliszyn. Anal Chem 68: 1521-1529, 1996.
40. JJ Langenfeld, SB Hawthorne, and DJ Miller. Anal Chem 68: 144-155, 1996.
41. L Pan, M Adams, and J Pawliszyn. Anal Chem 67: 4396-4403, 1995.
42. S Tutschku, S Mothes, and R Wennrich. Fresenius' J Anal Chem 354: 587-591, 1996.
43. T Górecki and J Pawliszyn. Anal Chem 68: 3008-3914, 1996.
44. DL Heglund and DC Tilotta. Environ Sci Technol 30: 1212-1219, 1996.
45. BL Wittkamp and DC Tilotta. Ana. Chem 67: 600-605, 1995.
46. AL Nguyen and JHT Luong. Anal Chem 69: 1726-1731, 1997.
47. Z Zhang and J Pawliszyn. Anal Chem 67: 34-43, 1995.
48. S Motlagh and J Pawliszyn. Anal Chem 284: 265-273, 1993.
49. M Chai and YZ Tang. Int J Environ Anal Chem In press.
50. T Górecki and J Pawliszyn. Anal Chem 67: 3265-3274, 1995.
51. M Chai, C Arthur, J Pawliszyn, R Belardi, and K Pratt. Analyst 118: 1501-1505, 1993.
52. CL Arthur, DW Potter, KD Buchholz, S Motlagh, and J Pawliszyn. LC/GC, 10: 657-681, 1992.
53. J Hinshaw. LC/GC, 7: 904, 1989.
54. R Shirey, L Nolan, and R Mindrup. The Reporter, Supelco, Bellefonte, PA, Vol. 15, No. 2: 2 1996.
55. LP Sarna, GR Webster, MR Friesen-Fischer, and RS Ranjan. J Chromatorg A 677: 201-205, 1994.

6

Food and Flavor Applications

Xiaogen Yang
Givaudan Roure Corporation, Cincinnati, Ohio

Terry L. Peppard *
Givaudan Roure Corporation, Clifton, New Jersey

INTRODUCTION

Research intended for understanding, designing, and improving the flavor of foods, beverages, or other flavored consumer products almost inevitably involves detailed chemical analysis of the finished product or raw materials used in manufacturing. Such chemical analyses normally comprise the following steps: (1) extraction and concentration of flavoring materials, (2) instrumental analysis of the resulting extract, and (3) interpretation of the results obtained. Most products contain trace levels of numerous flavor components, typically in the ppb to ppm concentration range, dispersed in a highly complex and often inhomogeneous matrix of carbohydrates, proteins, lipids, water, salts, etc. To overcome interferences from sample matrices and to enhance sensitivity of analytical methods, preconcentration and clean-up steps are in most cases unavoidable. A number of commonly employed techniques exist for extracting and concentrating flavor components from the product, including: solvent (liquid/liquid) extraction, static headspace sampling, dynamic headspace (Purge and Trap) sampling, simultaneous (Likens-Nickerson) distillation/extraction, supercritical fluid extraction (SFE), etc. Recently, Solid Phase Microextraction (SPME) has been applied for flavor analysis.

Current affiliation: Robertet Flavors, Inc., Piscataway, New Jersey

No one individual method is suitable for all purposes because of the complexity of food samples, the wide variation of flavor composition, and different analytical objectives. Each of the preconcentration techniques has its own advantages and disadvantages, and these methods are generally complementary to each other. To choose an optimal sample preparation method, one should consider the analytical objectives to be achieved, the type of product being analyzed, i.e. the product matrix (wet or dry, fat content, presence of browning or Maillard reaction precursors, continuous or particulate nature, etc.) and the type of flavor constituents present (volatility, water- and fat-solubility, tendency to oxidize and form acetals, or to be hydrolyzed, etc.). On the one hand, many flavor components tend to be relatively apolar and volatile in nature (e.g., terpene hydrocarbons, lipid-derived aldehydes, sulfur compounds, etc.). On the other hand, food matrices are frequently more polar and nonvolatile (e.g., water, proteins, carbohydrates, etc.). Therefore, SPME, which employs a fiber with a relatively apolar coating, provides a good method for preconcentrating flavor components, as well as certain preservatives, pesticides, and other contaminants, from many food, beverage, and related systems. SPME may work best by headspace or by liquid sampling depending on the nature of the matrix and the volatility of flavor components. In addition, SPME is relatively simple, fast, solvent-free, and inexpensive to carry out. Although SPME is no more a universal panacea than any of the other techniques listed above, it can, in many cases, provide a complementary or even superior alternative method. In this chapter, we will discuss the technical characteristics of SPME and review its application in the area of flavor analysis, specifically: foods, beverages, ingredients, botanicals, and additives or contaminants found therein.

GENERAL CONSIDERATIONS FOR SPME IN FLAVOR ANALYSIS

Selectivity and Sensitivity

The majority of flavor components are, as mentioned above, rather apolar in nature compared to the food matrix, itself. Commercially available SPME fibers, which include those coated with: (1) poly(dimethylsiloxane) (PDMS), (2) polyacrylate, (3) a poly(dimethylsiloxane)/divinylbenzene blend, and (4) a Carbowax/divinyl-benzene blend, are generally rather apolar or slightly polar. Accordingly, they can be used for preconcentrating flavor components from foods and related matrices. The selectivity of SPME fibers provides the basis for effective isolation of apolar trace compounds from a polar matrix. However, flavor components differ widely in polarity. Therefore, distribution coefficients of flavor components between an SPME fiber and a given food matrix can be expected to vary greatly.

Polarity and Acid/Base Dissociation Effects

We studied the adsorption behavior of 31 common flavor components from aqueous solution on SPME fibers coated with poly(dimethylsiloxane) and polyacrylate, respectively [1,2]. Table 6.1 shows the relative degree of adsorption measured for these flavor components; moreover, compounds were adsorbed on both fibers with dramatically different recovery rates. Although no apparent adsorption preference for particular compound classes was noted, the selectivity of SPME for flavor components is obvious. Table 6.1 indicates that adsorption selectivity is determined by polarity and molecular size under given experimental conditions.

Polarity of analytes can be used as a guide to the likely preconcentration efficiency in SPME. Relatively polar compounds, for example, ethyl vanillin, vanillin, furaneol, and furfural, which are usually difficult to isolate quantitatively from food samples using conventional methods like solvent extraction and Purge and Trap sampling, also have low affinity for the hydrophobic SPME fiber surface. Esterification of alcohols reduces the polarity and increases the molecular size of such analytes, and therefore, enhances the sensitivity of methods based on SPME.

As a general rule, if a molecule can be adsorbed on a surface, the larger the molecule is, the stronger it will be adsorbed under given conditions. This is also true for adsorption on SPME fibers, e.g., adsorption of γ-lactones increases with increasing molecular size. In this case, however, polarity also decreases with increasing length of the lactone ring side chain. The effect of molecular size on adsorption can also be seen in the example of aromatic aldehydes. The larger the molecule is, the stronger it will be adsorbed to a surface under given conditions.

Acids are common flavor ingredients, as well as food and beverage preservatives. Pan et al. determined fatty acids in water and air using a derivatization/SPME method [3,4]. The detection limits achieved with this method when analyzing short chain fatty acids were reported to be 1–4 orders of magnitude lower than those obtained using direct SPME. In general, acids can be present in food and beverage products in amounts ranging from trace (ppb) to low percentage levels. They can be the analytes of interest, or they can be interfering components in a sample. Fortunately, adsorption of acids on SPME fibers can easily be controlled by adjusting the pH of a liquid sample. The selective adsorption of acids on SPME fibers is easily understood from the relationship between adsorption and polarity. Dissociated and undissociated forms of acids coexist in aqueous solution, and the distribution of acid species is determined by the pKa value of the acid and the pH of the solution. The polarity of these two forms is very different. The ionic (dissociated) form is too hydrophilic (large hydration enthalpy) to be adsorbed on the hydrophobic surface of the fiber; thus only the neutral (undissociated) species are adsorbed on SPME fibers [1].

Table 6.1 Relative Adsorption on SPME Fibers.

Component	Polyacrylate (85μm)	PDMS (100μm)
Terpenes		
α-pinene	1.2	7.3
limonene	3.1	15.1
Esters		
ethyl acetate	0.3	0.5
iso-amyl formate	3.1	14.9
ethyl hexanoate	24.9	52.9
cis-3-hexenyl acetate	24.3	52.5
diethyl malonate	2.3	2.4
triacetin	0.8	0.7
Ketones		
heptanone	7.2	18.1
β-ionone	37.9	62.2
Lactones		
γ-hexalactone	0.9	0.9
γ-decalactone	44.6	42.0
γ-dodecalactone	59.8	73.3
Alcohols		
cis-3-hexenol	2.5	0.9
linalool	35.0	40.8
l-menthol	44.8	56.0
phenylethyl alcohol	4.2	0.8

Table 6.1 Continued.

Component	Polyacrylate (85μm)	PDMS (100μm)
Aldehydes		
furfural	0.7	0.4
benzaldehyde	7.3	3.4
neral	32.3	40.7
geranial	40.9	45.2
cinnamic aldehyde	17.9	5.9
Ether		
anethole	49.5	74.8
Acid		
hexanoic acid	5.4	0.7
Multifunctional		
guaiacol	1.2	3.0
eugenol	16.8	17.2
heliotropine	11.1	2.2
eugenyl acetate	58.5	64.5
triethyl citrate	1.4	0.6
ethyl vanillin	4.3	<0.1
vanillin	1.7	<0.1

The influence of pH on SPME adsorption for acetic acid is shown in Figure 6.1. As expected, the relative degrees of adsorption on poly(dimethylsiloxane) and polyacrylate fibers show trends that correspond to the distribution of undissociated species of acid with pH. Therefore, lowering the pH value of a sample solution increases an SPME method's sensitivity for acids, while raising the pH can reduce interferences from high concentrations of acidic components in an analysis. The longer the fatty acid chain, the stronger the adsorption.

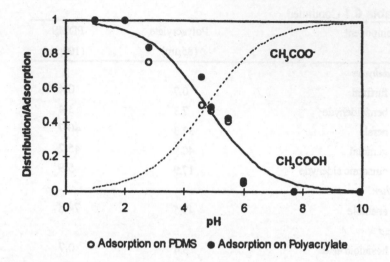

Figure 6.1 Distribution of acetic acid in aqueous solution and relative adsorption on SPME fibers.

Solubility Effect

A molecule has to be desolvated before it can be adsorbed from an aqueous solution. The enthalpy change after adsorption is the adsorption enthalpy minus the hydration enthalpy of the compound and the surface. Therefore, a smaller solvation enthalpy is more favorable for adsorption. In other words, the higher the solubility of a compound in a food or beverage matrix, the less it tends to be adsorbed on the SPME fiber. The relationship [1] between adsorption and solubility is shown in Figure 6.2. Despite the large variety of compound classes included, a clear trend can be seen: adsorption decreases with increasing solubility in aqueous solution. Solubility plays a crucial role in an adsorption system; this is true not only for aqueous solutions, but also for solutions in such media as vegetable oil, organic solvent/water mixtures, etc. Considering the solubility and concentration levels of analytes in a matrix can generally provide a guide as to the likely success of SPME for a given sample. In general, SPME behavior agrees with what would be expected from knowledge of adsorption and liquid/liquid partitioning. Therefore, the suitability of SPME for preconcentration of particular analytes in a given sample can be gauged according to experience gained from conventional analytical techniques.

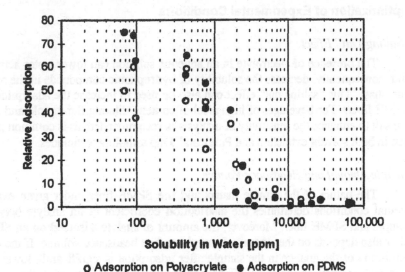

Figure 6.2 Adsorption *vs.* solubility (From Ref. 1).

Porous Layer SPME

The SPME technique is a single batch method; moreover, the surface area available is very limited (typically less than 10 mm^2). Therefore, to provide the basis of a sensitive method, an SPME fiber must be coated with a high degree of efficiency. So far, relatively few SPME fiber coatings have been successful, though the number is gradually increasing.

Pure silica, such as the uncoated SPME fused silica fiber, can also adsorb trace flavor components from botanicals and food products [5]. However, the silica fiber has a polar surface due to the presence of silanol groups; accordingly, it has a low affinity for hydrophobic compounds. Consequently, the sensitivity of SPME methods based on the use of an uncoated fused silica fiber will be lower than that achieved when employing fibers with various polymer coatings. When the silica surface is chemically modified with C_{18} or C_8, it becomes hydrophobic and can adsorb apolar compounds from a polar matrix, as occurs with reversed phase HPLC columns. Liu *et al.* [6] used a metal wire coated with bonded silica particles as an SPME fiber. Porous silica particles provide a much larger specific surface area than the polymers used to date. The authors reported that a fiber coated with a 30 μm C-8 porous layer provided eight times the sensitivity of a fiber coated with 100 μm poly(dimethylsiloxane). This approach seems promising in the development of improved fibers for highly sensitive flavor analysis.

Optimization of Experimental Conditions

"Salting-Out" Effect

The presence of electrolyte in an aqueous solution can lower water activity, and, consequently, decrease the solubility of hydrophobic compounds in the aqueous phase. This "salting out" effect can also be used to increase the adsorption on SPME fibers. For beverages of high alcohol content, diluting the sample and adding salt to enhance the sensitivity of analysis is recommended, as is common practice in both solvent extraction and Purge and Trap sampling techniques.

Sample Amount and Headspace Volume

The nature of the sample matrix and the SPME fiber under given experimental conditions determines the distribution coefficient of an analyte between sample and SPME fiber. Moreover, the amount of analyte adsorbed on an SPME fiber also depends on the amount of sample and its headspace volume. If the concentration of the analyte in the sample after adsorption is significantly lower, the amount adsorbed on the SPME fiber can be increased by increasing the sample amount, while the distribution coefficient remains constant. With increasing sample amount, the amount adsorbed also increases, at first, but then remains practically constant thereafter. For analysis of beverage flavors, several milliliters of sample is generally sufficient. For a solid or liquid sample, analytes distribute among the condensed phase(s), the adsorption phase and the gas phase. Because the total amount of analytes in the three phases is constant at equilibrium, lowering the headspace volume will increase the amount of the analytes in the other two phases, and therefore, increase the sensitivity of SPME-based methods. The effect of headspace volume is often most pronounced for volatile trace flavor components.

Adsorption and Desorption Rate

Adsorption rate during SPME sampling has been studied in a number of papers [6,7,8]. Sampling kinetics involves evaporation, dissolution, diffusion, adsorption, etc. The kinetic parameters are dependent not only on experimental conditions, but also on the physical properties of the sample matrix and analytes. The amount adsorbed on an SPME fiber increases with sampling time until reaching a plateau.

For liquid sampling, the adsorption rate of SPME is mainly determined by mass transfer and diffusion processes, and is thus accelerated by agitation [6,7]. The time needed to reach the plateau is typically around 10 minutes [8]. For headspace sampling, the SPME process involves evaporation of the analytes into the gas phase, and diffusion to the surface of the fiber. Adsorption from the gas phase onto the fiber surface is fast in comparison with that from a liquid phase because the diffusion rate is much greater in the former case. However, evaporation rates can be very different for various analytes. For compounds with lower vapor pressure, evaporation seems to be the rate determining step. For volatile components of apple, such as ethyl butanoate and propyl butanoate, the plateau of adsorption vs. time can be reached within 5 to 10 minutes, while accumulation on the fiber is still increasing after 90 minutes for less-volatile components such as hexyl butanoate, hexyl 2-methylbutanoate, and α-farnesene [9]. For hop oil analysis using SPME, a steady-state of adsorption had not been reached for caryophyllene and humulene even after a 6-hour sampling period [10]. In the case of orange juice analysis, required equilibration time for headspace sampling ranged from 5 to 120 minutes [11].

Continuing to sample until the adsorption equilibrium has been reached can achieve maximal sensitivity, as well as minimal variation, for an SPME-based method. However, long sampling times reduce the advantage of SPME as a rapid method. Fortunately, sampling to equilibrium in SPME is not necessary for many flavor analyses. Even for quantitative analysis, if the sampling time and conditions are accurately controlled, acceptable reproducibility can be achieved.

If the temperature of the GC injection port used for desorption was set at 200°C, carryover was observed in our lab for some major components of essential oils. At 250°C, however, Schäfer et al. reported that 95% of analytes adsorbed on an SPME fiber were desorbed in 1 second, while the desorption was quantitative after 10 seconds [12].

If SPME is used as a rapid screening method, optimization of sampling time and desorption time in the GC injection port is not necessary. We generally use 3 to 10 minutes for sampling, and 3 minutes for desorption.

Liquid vs. Headspace Sampling

Liquid sampling involves immersing the SPME fiber directly into the liquid phase of a sample, whereupon trace components are enriched on the fiber's surface. This technique is suitable only for "clean" food and beverage samples that do not contain large amounts of nonvolatile components, as these may be adsorbed to the fiber surface as well. Some beverages are suitable for using the liquid sampling technique. Empirically, if the concentrate of a liquid sample produced by solvent extraction can be directly injected into a gas chromatograph without seriously contaminating the injection port and capillary column, then SPME can probably also be used with a liquid sampling technique without causing undue contamination problems.

For liquid and solid samples, three phases are generally involved during the SPME process. If the property change of the adsorption surface in gas and liquid phases can be ignored, the distribution of an analyte in the three phases at equilibrium will not change, irrespective of where the SPME fiber is placed (liquid or headspace). This means that the amount of an analyte adsorbed on the SPME fiber should be the same for both liquid and headspace sampling, *providing equilibrium has been reached* [2]. In a number of cases, the amount of material adsorbed seemed to be similar for liquid and headspace procedures. For instance, when

SPME was used for analysis of pyrazines in model reaction systems, sensitivity of liquid and headspace sampling was comparable [13]. This was also observed by Coleman in the analysis of Maillard reaction products [8].

Liquid and headspace sampling methods differ in kinetics, not only due to possible differences in rates of evaporation, dissolution, and diffusion in gas and liquid phases, but also due to the concentration difference between the liquid phase and its headspace. These factors make SPME liquid and headspace sampling methods very different in typical nonequilibrium situations. In general, the adsorption rate is higher when the concentration of analyte is higher. Therefore, if an analyte exists predominantly in the liquid phase, an SPME liquid sampling method is more sensitive than a headspace sampling method (for a given sampling time) and *vice versa*. This observation leads to the selection of headspace sampling as the best means of separating volatile from less-volatile compounds.

To successfully enrich trace flavor components on an SPME fiber, the analytes must have significantly different polarity and/or vapor pressure from those of the majority of components in the sample. SPME liquid sampling utilizes distribution differences to enrich trace components from the sample matrix. In addition to the polarity selectivity factor, headspace sampling also uses volatility differences to "clean-up" the sample. Although the food matrix itself is generally nonvolatile in nature, headspace sampling SPME typically has higher application potential because most flavor compounds are relatively volatile.

Rules of Thumb

Summarized below are some rules of thumb for gauging the suitability of SPME for a given flavor analysis and for optimizating experimental conditions.

1. Sensitivity decreases with increasing analyte solubility in a matrix.
2. For a given class of compounds, sensitivity increases with molecular size.
3. Addition of electrolyte in sample solution generally increases sensitivity.
4. Larger sample size gives higher sensitivity, until a plateau is reached.
5. Smaller headspace volume gives higher sensitivity.
6. Differences in polarity and volatility between the analytes and matrix provide the basis for SPME preconcentration. Liquid sampling utilizes the polarity differences, whereas headspace sampling also uses volatility differences.

APPLICATIONS

Beverages

The first reported application of SPME for beverage analysis appears to have been determining caffeine using an uncoated silica fiber [5]. A liquid sampling period of 5 minutes was applied for brewed coffee, tea, and carbonated cola soft drink, followed by GC-MS analysis. For quantitation, the authors used isotopically-labeled caffeine as the internal standard that enabled a relative standard deviation of about 5% to be achieved. Major flavor components such as terpenes, esters, and acids in tea, several carbonated soft drinks, fruit juices, and alcoholic beverages were also detected using this method. However, as the authors indicated, SPME is very difficult to use for quantitation in multicomponent analyses without the addition of appropriate internal standards and/or knowledge of analyte distribution coefficients between the sample matrix and SPME fiber.

SPME liquid sampling was compared with solvent extraction for analyzing a fruit juice-based beverage (Figure 6.3). Most of the 25 flavor components extracted into methylene chloride were also concentrated on the SPME fiber coated with 100 μm poly(dimethylsiloxane), although the component ratios observed were quite different [2]. SPME was found to be more sensitive for most esters, terpenoids, and lactones.

To characterize vodkas of different origin and brand, Ng and coworkers applied SPME to enrich certain of the ethyl esters of C_8 to C_{18} fatty acids, at the ppb level, from products containing 40% ethanol [14]. Ethyl hexadecanoate was used as the internal standard. A sampling time of 1 hour was used even though the adsorption equilibrium had not been reached after two hours of liquid sampling. Standard deviations of 1 to 20% were achieved for the concentration range 0.1–32 ppb. The ethyl ester profile could be used as an indicator of raw materials utilized in the fermentation and other production procedures employed in the manufacture of vodkas.

Elmore et al. compared SPME headspace sampling with both poly (dimethylsiloxane)- and polyacrylate-coated fibers to dynamic headspace concentration on Tenax TA for analyzing aroma volatiles in cola and diet cola [15]. The Tenax method gave a higher yield of volatiles (but similar profile) to SPME using poly(dimethylsiloxane). Although SPME using a polyacrylate fiber yielded fewer components than the other two methods, these did include two polar volatiles not otherwise seen. Reproducibilities for the two techniques were comparable. Qualitative differences between the two cola types were small, though the regular cola yielded approximately four times the total quantity of volatiles.

SPME followed by GC analysis with sulfur chemiluminescent detection (SCD) was applied for qualitative detection of trace sulfur components in a carbonated beverage in the authors' lab. An SPME fiber coated with poly(dimethylsiloxane) seemed to have strong affinity for sulfur compounds that were suspected of being the cause of an off-odor. Before liquid sampling, salt was added and carbon dioxide was removed by placing the sample in an ultrasonic bath for a short period until no more gas was evolved. Sensitivity was comparable to that obtained using a solvent extraction method.

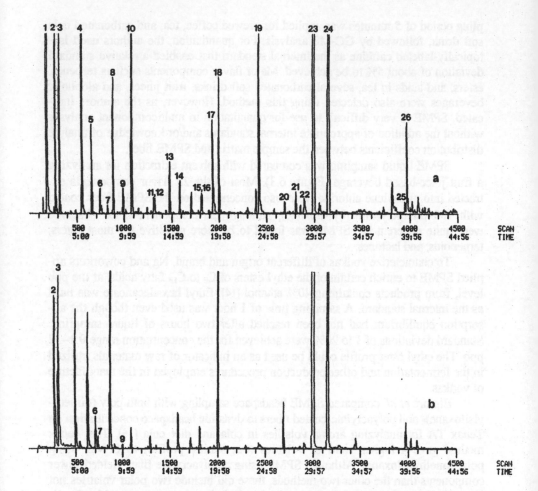

Figure 6.3 Gas chromatograms of fruit juice beverage: (a) solvent extraction with methylene chloride; (b) SPME liquid sampling (100 μm polydimethylsiloxane). 1. dichloromethane; 2. ethyl butyrate; 3. ethyl isovalerate; 4. limonene; 5. ethyl hexanoate; 6. isoamyl butyrate; 7. hexanyl acetate; 8. *cis*-3-hexenyl acetate; 9. hexanol; 10. *cis*-3-hexenol; 11. *cis*-3-hexenyl butyrate; 12. furfural; 13. benzaldehyde; 14. linalool; 15. β-terpineol; 16. butyric acid; 17. 2-methylbutyric acid; 18. α-terpineol; 19. hexanoic acid; 20. *cis*-methyl cinnamate; 21. 1-(2-furyl)-2-hydroxyethanone; 22. furaneol; 23. *trans*-methyl cinnamate; 24. γ-decalactone; 25. dodecanoic acid; 26. hydroxymethylfurfural. (From Ref. 30.)

Fruits

Flavor components of fresh strawberry juice were analyzed using an SPME fiber coated with poly(dimethylsiloxane) [16]. However, no satisfactory results could be achieved using direct liquid sampling due to the presence of large amounts of nonvolatile components. These competed with the trace flavor components for adsorption on the SPME fiber surface. Adsorbed nonvolatile components also interfered with the subsequent GC analysis, causing noisy baselines and producing interference peaks. Additionally, the SPME fiber was difficult to clean. As a result, SPME headspace sampling turned out to be more suitable for this kind of sample. For comparison, solvent extraction was applied to the sample, and approximately 150 volatile compounds were subsequently identified. SPME enriched major volatile components such as butyrates and hexanoates; although, it performed rather poorly from the sensitivity standpoint for γ-decalactone and methyl anthranilate, and especially for the key flavor components of strawberry aroma, namely, furaneol and mesifurane. SPME for highly volatile components yielded low reproducibility (13–26% relative standard deviation). In contrast, better reproducibility was achieved in determining trans-2-hexenal, methyl caproate, γ-decalactone, and methyl anthranilate. SPME could be used as a screening method for a juice production process.

SPME headspace sampling was used to monitor flavor changes in banana during the freeze drying process [17]. Volatile components were completely lost after 5 to 8 hours of freeze drying, although less-volatile compounds such as isoamyl acetate, isobutyl butyrate, butyl butyrate, isoamyl butyrate, and isoamyl alcohol decreased to 8–25% of initial concentration and then remained constant. The results obtained using SPME were comparable to those obtained by steam distillation/extraction analysis. For comparison, conventional headspace sampling was also used for the analysis; SPME headspace sampling was more sensitive for less volatile compounds than was conventional headspace analysis.

Matich et al. evaluated an SPME headspace sampling technique to determine additional formation of flavor components in apple during cool storage [9]. They found the evaporation of volatile components from the sample to gas phase to be the rate determining step in the SPME sampling process; furthermore, the evaporation rates greatly differed. To illustrate, equilibrium for SPME adsorption was reached within 5 minutes for compounds with lower molecular weights (<200 MW). By contrast, equilibrium was not reached even after 90 minutes for compounds of higher molecular weight. Increasing gas movement can help increase the adsorption rate. In comparison with the results obtained using a dynamic headspace (Purge and Trap) method, the component ratios obtained were very different. The SPME fiber coated with poly(dimethylsiloxane) preferentially adsorbs volatile components of higher molecular weight.

A rapid analytical method was established for the analysis of volatile apple components [18]. This method was a combination of SPME, fast GC, and time-of-flight MS. After a 4 to 6 hour equilibration period, the headspace above apples was sampled using an SPME fiber coated with 100 μm poly(dimethylsiloxane) for up to 24 minutes. The GC run was completed in several minutes, and 29 volatile components of apple were successfully identified. Linear response was found in the ppb to ppm concentration range.

Spices and Other Botanicals

SPME provides an excellent tool for the analysis of the volatile components of botanicals. The conventional methods for essential oil analysis involve steam distillation, direct thermal desorption, or headspace sampling. Steam distillation is time-consuming and prone to artifact formation. Direct thermal desorption can also lead to artifacts, headspace sampling can provide insufficient sensitivity, and both methods can introduce water into GC systems, interfering with subsequent analysis. A number of other analytical methods have also been developed for essential oil analysis, including CO_2 extraction, solid-phase extraction, and vacuum steam distillation. All of these are multi-step methods which often require sample clean-up before GC analysis.

We have used SPME to analyze essential oil composition to gauge the quality of spices. Figures 6.4 and 6.5 show results obtained for cassia bark and basil, respectively, using static headspace GC, employing direct liquid injection of essential oil obtained by steam distillation, and headspace SPME GC. The SPME fiber coated with poly(dimethylsiloxane) showed high adsorptive capacity for essential oil components. All major compounds that were detected and identified in the steam distilled and static headspace-sampled essential oils were also adsorbed by the SPME fiber. However, the ratios of detected essential oil components varied between methods, due to volatility differences and because of the selectivity of the SPME fiber. Static headspace GC is more sensitive for early-eluting peaks; while headspace SPME GC is more sensitive for later-eluting peaks. In addition to the advantages of speed, simplicity, and cost, SPME causes less artifacts than steam distillation. For example, benzaldehyde (peak 1 in Figure 6.4) can be produced as an artifact from cinnamic aldehyde (retro-aldol condensation) during the steam distillation process. Under SPME sampling conditions, such reactions are unlikely to occur. On the other hand, certain compounds that are not stable under hydrolytic conditions (therefore, are often absent in essential oils obtained by steam distillation) may be detected using SPME. The breakdown and/or rearrangement of linalyl acetate to linalool and a series of other terpene alcohols, esters, and hydrocarbons during steam distillation of hop essential oil is a prime example of this phenomenon [19]. Another example from the authors' laboratory is provided by 1'-acetoxychavicol, which is one of the major components of hexane extract of galangal root, but which cannot be detected in steam distillates prepared at 100°C and ambient pressure.

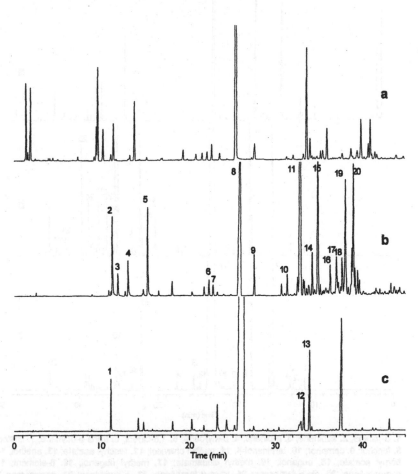

1. benzaldehyde; 2. α-pinene; 3. comphene; 4. β-pinene; 5. 1,8-cineopl; 6. terpinen-4-ol; 7. α-terpineol; 8. cinnamic aldehyde; 9. bornyl acetate; 10. α-cubebene; 11. α-copaene; 12. coumarin; 13. cinnamyl acetate; 14. α-bergamottene; 15. β-caryophyllene; 16. α-humulene; 17. germacrene D; 18. β-selinene; 19. α-muurolene; 20. δ-cadinene

Figure 6.4 Gas chromatograms of cassia bark volatiles: (a) conventional headspace sampling; (b) SPME headspace sampling; (c) direct injection of essential oil.

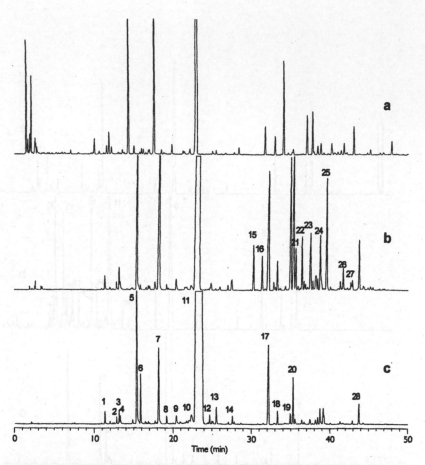

1. α-pinene; **2.** sabinene; **3.** β-pinene; **4.** β-myrcene; **5.** 1,8-cineol; **6.** trans-b-ocimene; **7.** linalool; **8.** fenchol; **9.** camphor; **10.** terpinenol-4; **11.** methyl chavicol; **12.** fenchyl acetate; **13.** anethol; **14.** bornyl acetate; **15.** eugenol; **16.** methyl cinnamate; **17.** methyl eugenol; **18.** β-element; **19.** caryophyllene; **20.** cis-α-farnesene; **21.** trans-β-farnesene; **22.** α-humulene; **23.** germacrene D; **24.** γ-cadinene; **25.** γ-bisabolene; **26.** caryophyllene oxide; **27.** humulene epoxide II; **28.** trans-cadinol

Figure 6.5 Gas chromatograms of basil volatiles: (a) conventional headspace sampling; (b) SPME headspace sampling; (c) direct injection of essential oil.

Coleman and Lawrence evaluated four different GC-based analytical methods for their ability to provide qualitative and quantitative information on the essential oil of Virginia cedarwood (*Juniperus virginiana* L.) [20]. The methods were SPME employing headspace sampling and a fiber coated with poly-(dimethylsiloxane), both static and dynamic (Purge and Trap) headspace analysis, and direct injection of essential oil. They concluded that each method had its own strengths and limitations, and that the choice of which method to use ultimately depends on the specific objectives of the work. For example, direct injection of an essential oil gives the most reliable quantitative information on the oil *per se*. If information concerning the components responsible for the aroma of an essential oil is desired, however, then the use of static headspace analysis may be preferable. They further concluded that although both SPME and dynamic headspace sampling generally give similar information, fiber exposure time (in SPME) and sweep time (in Purge and Trap sampling) have a critical influence on the results obtained.

SPME headspace sampling at both room temperature and 80°C was used in analyzing the flavor of Pico white truffles (*Tuber magnatum*) and black Perigord truffles (*Tuber melanosporum*) [21]. A series of sulfur compounds was identified as dimethyl sulfide, dimethyl disulfide, bis(methylthio)methane, dimethyl trisulfide, 1,2,4-trithiolane, methyl(methylthio)methyl disulfide and tris(methylthio)-methane. Only two major non-sulfur components, 2-butanone and 2-butanol, were enriched on the SPME fiber coated with poly(dimethylsiloxane). Accordingly, SPME appears to have good selectivity for sulfur compounds. For comparison, direct thermal desorption was also used for the analyses. Although thermal desorption was more sensitive for most of the volatile components, SPME sampling effectively avoided introduction of water into the GC column. A disadvantage of SPME was its selective adsorption of less polar compounds, which resulted in a distorted composition profile.

Conifer Needles (*Pinus peuce* Griseb., *Pinus strobus*, *Pinus schwerinii*, and *Pinus monticola*) were analyzed using SPME GC [12]. Sampling temperature has two adverse effects on the SPME process: sampling time required for reaching an adsorption steady state is longer at lower temperatures due to lower diffusion rate, and any matrix effects are stronger at lower temperature. On the other hand, adsorption of analytes on the SPME fiber decreases with increasing temperature. For monoterpenes, desorption at 250°C was practically quantitative after 10 seconds with more than 95% of the examined compounds already desorbed after 1 second. As various analytes have different distribution coefficients, composition determined by SPME GC does not necessarily represent the true composition of the sample headspace. The monoterpene composition of conifer needles that was determined by using SPME headspace sampling was compared to that obtained using a simultaneous distillation/extraction method. SPME was considered applicable for these analyses. However, lower recoveries for compounds with higher boiling points was noted, which is caused by incomplete vaporization of such substances into the gas phase. Reduction of analyte/matrix interactions was very difficult. Standard addition was not practical and only slight improvement was achieved by matrix modification (addition of 10 mL methanol to the sample) and by increasing headspace temperature from 50 to 90°C. Despite the difficulties of quantitative analysis by SPME, the results obtained by SPME GC can be used as "fingerprints" of botanical samples in order to identify plant species and varieties.

To classify samples by botanical origin, profiles of cinnamon and cassia were determined by SPME headspace sampling followed by GC-MS analysis [22]. A powdered cinnamon sample of 10 mg was preheated for 15 minutes at 70°C before SPME sampling for 5 minutes. The SPME fiber coated with 7 μm poly(dimethylsiloxane) gave lower recovery of semivolatile components than did one coated with 100 μm. Under these experimental conditions, the SPME fiber coated with 100 μm poly(dimethylsiloxane) apparently exhibited higher adsorption of linalool, camphor, β-caryophyllene, cinnamyl acetate, α-humulene, and coumarin. On the other hand, the fiber coated with 85 μm polyacrylate showed higher adsorption of benzaldehyde, limonene and cinnamic aldehyde. The actual number of adsorbed compounds was almost the same for both fibers, and either one could be used for this analysis. Although the headspace profile obtained by SPME using a fiber coated with 100 μm poly(dimethylsiloxane) was significantly different from that obtained by supercritical fluid extraction (acetonitrile as modifier), reasonable correlation was observed nonetheless for the major semivolatile compounds (r^2 ranged from 0.87 to 1.00). Higher concentrations of terpenes were obtained by SPME headspace sampling. The key components that could be used to distinguish between true cinnamon and cassia were eugenol and benzyl benzoate (present in cinnamon, but not in cassia) and coumarin and δ-cadinene (present in cassia but absent or at low concentration in true cinnamon). The four components in cinnamon can easily be analyzed using SPME headspace sampling. Therefore, SPME provides an effective, rapid, and simple method to classify the origin of cinnamon.

The ratio of humulene to caryophyllene (H/C) in hops (*Humulus lupulus*) provides a useful predictor of hop aroma in beer. SPME was applied to determine this characteristic parameter [10]. Unlike the monoterpenes, which needed only a few minutes or less to reach an SPME adsorption plateau [12], the sampling-steady state for sesquiterpenes humulene and caryophyllene in hops was not reached, even after 6 hours at 50°C [10]. Nonetheless, an accurate H/C ratio could be obtained after 4 hours sampling, and a relative standard deviation of 2.3% was achieved for this determination. The SPME results agreed with those obtained using steam distillation and pentane extraction. SPME was thus found to be a good alternative screening tool for use in selecting hops for breeding purposes.

Processed Foods and Maillard Reaction Products

Ground espresso-roast coffee was analyzed by SPME headspace sampling [2]. Preferential adsorption on an SPME fiber coated with poly(dimethylsiloxane) was also observed for semivolatile components.

Most flavor components are readily soluble in vegetable oils and fats. To isolate flavor components from such systems, elevated sampling temperatures are needed. At 160°C, the major flavor components of butter flavor, i.e., diacetyl, δ-decalactone, and δ-dodecalactone, were detected using SPME headspace sampling [2].

Pyrazines and their formation mechanism(s) in food are important topics in flavor science. To evaluate the use of SPME for the analysis of pyrazines, a model system was established composed of a heated aqueous solution of glucose, glycine, and sodium hydroxide [13]. Analytical results obtained by SPME headspace sampling from this model reaction system were comparable to those obtained by liquid sampling and solvent extraction, most likely due to similarities in polarity and volatility of the pyrazine derivatives involved. SPME headspace sampling did, however, appear more sensitive for volatile components.

A complex mixture of volatile and semivolatile compounds is produced from the chemical reactions of proteins with reducing sugars. These Maillard reaction products, which include pyrazines, thiazoles, furans, aldehydes, and pyridines, are important sources of desirable food aromas as well as off-flavors, and are therefore important subjects for study in food and flavor analysis. Coleman evaluated SPME for the analysis of Maillard reaction products and substances derived from thermal breakdown of sugars [8,23]. Sixteen compounds originating from Maillard reactions at 50 ppm in aqueous solution were preconcentrated using an SPME fiber coated with 100 μm poly(dimethylsiloxane), and subsequently analyzed by GC-MS [8]. The detection limit for each compound was approximately 1–2 ng; moreover, the SPME fiber exhibited high selectivity. The amount of analyte adsorbed on the fiber increased with degree of alkyl substitution, which reduces molecular polarity and increases molecular size. The presence of alkoxy substituents also enhanced the degree of adsorption, although carbonyl substitution had the opposite effect. In several cases, the amounts adsorbed by headspace and liquid sampling were comparable. Addition of sodium chloride increased adsorption for all compounds tested. Adsorption interference was observed in the presence of other analytes. For example, the amount of pyridine adsorbed on the SPME fiber was reduced by about 50% in the presence other pyridine derivatives. An SPME fiber coated with Carbowax-divinylbenzene, which is more polar than one coated with Poly(dimethylsiloxane), was also used to analyze Maillard reaction products [23]. This polar fiber was found to be more sensitive for most of the compounds analyzed, but was slightly less sensitive for compounds such as 4-ethylpyridine, 5-ethyl-2-methylpyridine, and trimethylthiazole. Both apolar and polar SPME fibers appear well suited for the analysis of Maillard reaction products in aqueous solutions. The detection limit lies in the ppb concentration range, and linear adsorption can be obtained from 5 to 500 ppb. For quantitative analysis, however, the competitive adsorption of analytes must be taken into consideration.

Food Additives and Contaminants

Page and coworkers analyzed volatile halogenated contaminants in various foods and beverages using SPME headspace sampling and a fiber coated with poly(dimethylsiloxane) [24]. Standard addition was used for quantitation. Chloroform, bromodichloromethane, and chlorodibromomethane (by-products of water chlorination) were found in some beverages at concentration levels ranging from 0.2 to 15 ppb. Dichloromethane was found in flour products and in coffees and teas that were decaffeinated using dichloromethane. 1,1,1-Trichloroethane was found as a packaging contaminant at a level of *ca.* 200 ppb in a biscuit mix. Detection limits in a water model solution of *ca.* 1.5 µg/kg for vinyl chlorine and <0.005 µg/kg for tri- to hexachlorobenzenes were achieved using an electrolytic conductivity detector. SPME headspace sampling was more sensitive for less volatile components than conventional headspace analysis, and *vice versa.* Food matrix effects for SPME were also observed. Lipid material in a sample greatly reduces the SPME sensitivity, especially for less-volatile components. For example, the presence of 50 mg lipids in 1 g food sample decreased the adsorbed amount on SPME GC analysis by 50-99%.

Residual Procymidone, a fungicide widely used on wine grapes, can be determined in wine at concentrations ranging from a few ppb to a few hundred ppb. An SPME fiber coated with 100 µm poly(dimethylsiloxane) was used to preconcentrate procymidone from wine [25]. In combination with selected ion monitoring, a detection limit of 0.5 ppb was achieved. The analytical results obtained using SPME were in agreement with those measured using an enzyme-linked immunosorbent assay (ELISA).

Maltol and ethyl maltol are common flavoring ingredients in a variety of food products. Maltol has been found to be present in natural products, although ethyl maltol, to date, has not. An analytical method based on SPME was developed to analyze the concentration of maltol and ethyl maltol in coffee, beverages, chewing gums, and potato chips [26]. Due to the acidity of these two compounds, their adsorption on an SPME fiber coated with 100 µm poly(dimethylsiloxane) was pH-dependent: the lower the pH, the higher the degree of adsorption. The addition of salt enhanced the adsorption. For solid samples, microwave-assisted extraction was carried out prior to SPME sampling. In combination with single ion monitoring (SIM), the GC-MS detection limit was found to be 10 ppb for maltol and 2 ppb for ethyl maltol; furthermore, a relative standard deviation of 13% was achieved.

SPME headspace sampling coupled with GC-MS was used by Clark and Bunch for qualitative and quantitative determination of a series of flavor additives on spiked Kentucky tobacco [27]. Flavor additives included in the study were menthol, anethole, benzaldehyde, and tetramethylpyrazine, as well as mandarin orange, nutmeg, and sweet fennel oils. SPME fibers evaluated were those coated with 100 µm poly(dimethylsiloxane), 65 µm polyacrylate, 65 µm poly(dimethylsiloxane)/divinylbenzene, and 65 µm Carbowax/divinylbenzene, each resulting in a significantly different component profile. Methods were optimized by varying experimental parameters such as time, temperature, presence of salt, etc. Limits of detection for 31 typical tobacco flavorants were mostly below 500 ng/g, but ranged from 15 ng/g to *ca.* 6 μg/g. The four individual flavor substances listed above yielded recoveries ranging from 72 to 128% and relative standard deviations ranging from 5.3 to 17.6%. Major components in the various essential oils exhibited linear relationships with oil concentration.

A taint problem in wine, due to the presence of 2,4,6-trichloroanisole derived from cork, was solved using GC-MS following SPME with headspace sampling onto a poly(dimethylsiloxane)-coated fiber [28]. The method's 2.9 ng/L limit of detection was considered low enough to enable detection of the off-flavor in wine at or slightly below 2,4,6-trichloroanisole's reported flavor threshold of 4 to 50 ng/L. An SPME liquid sampling method was also tested, but found to be inferior in sensitivity and to lead, in any case, to contamination problems.

QUANTITATIVE ANALYSIS

A high recovery rate is generally required for an accurate and precise quantitative analysis using external calibration. Because SPME as a preconcentration procedure is a single batch method, quantitative recovery in most cases is very difficult, if not impossible, to accomplish. The results of beverage flavor analysis by SPME and liquid/liquid extraction are compared in Figure 6.3. Although most of the flavor components extracted by methylene chloride were also adsorbed on the SPME fiber, the relative recoveries achieved were very different. Whereas the phase ratio, sample amount, concentration factor, and number of extraction steps employed in liquid/liquid extraction are easy to change to suit analytical needs regarding sensitivity and selectivity, SPME is more restrictive.

In practice, to compensate for substantially incomplete analyte recoveries, an internal standard that possesses similar physical and chemical properties is often used. However, choosing an appropriate compound as an internal standard is difficult due to the high selectivity of SPME fibers. Thus, standard addition or stable isotope dilution techniques are more suited to SPME-based analytical methods. The determination of caffeine in beverages demonstrated the use of an isotope dilution technique in quantitative analysis by SPME [5]. In our laboratory, SPME with standard addition has been successfully applied for determining a number of contaminants in beverages. However, the application of standard addition and isotope dilution methods are limited to liquid samples, where thorough mixing and complete homogeneity can be ensured, and of course, analytes for which the corresponding standard compounds or isotopically labeled analogues are available. Furthermore, the standard addition method requires multiple analyses to be carried out for a single sample, thereby diminishing some of the advantages of SPME as a rapid and simple technique. Standard addition and isotope dilution procedures are also not very practical in the case of multicomponent analyses.

Ai described the use of SPME for quantitative analysis specifically in non-equilibrium situations [29,30]; sampling times were much shorter than required for reaching adsorption equilibrium, but plots of adsorbed amount *vs.* initial concentration of analyte nonetheless showed excellent linearity.

Quantitative analysis can be carried out if the distribution coefficients of analytes between an SPME fiber and the sample matrix are known. Nevertheless, the method can only be extended to other systems if the distribution coefficients remain unchanged. This idea could be useful for gas phase analysis, where the matrix effect is generally negligible, while "real world" samples are more problematic. An interesting approach to quantitation was discussed by Schäfer *et al* [12]. Under given experimental conditions, two factors determine the amount of analyte adsorbed on the SPME fiber: the analyte's vapor pressure and its distribution between gas and adsorption phases, i.e., the two distribution coefficients applying among the three phases generally present in food systems. Quantitation is possible if the two distribution coefficients are known, although unfortunately, these data are rarely available for most foods or botanical samples. However, this problem can be overcome using the so-called "full evaporation technique"[31] that initially forces all analytes into the gas phase. Total evaporation of analytes into the gas phase not only eliminates sample matrix effects, but also simplifies the analytical system into two phases: the gas phase and the SPME adsorption phase. At this point, one only need know the distribution coefficient of adsorption for quantitation to be possible.

The retention time of an analyte in GC is proportional to the distribution coefficient of that analyte between carrier gas phase and column stationary phase. Therefore, the relative distribution coefficient of an analyte between the gas phase (the headspace above a food) and the SPME adsorption phase can be estimated from its Kovats' retention index on a GC column having the same coating as the fiber. A very good correlation exists between the relative distribution factor and Kovats' index for monoterpenes [12]. The authors also mentioned using an homologous series of n-alkanes as the basis of a standard system, which is similar to the retention index system of Kovats. This approach can be applied not only in the analysis of botanicals, but also in other headspace sampling procedures. An external standard can be used to determine the response factor of the GC detector. Martos *et al.* further explored this approach for the analysis of petroleum hydrocarbons in air [32]. A very good correlation ($r^2 = 0.9999$) of the distribution coefficients between the SPME fiber coating and headspace was obtained with corresponding GC retention indices. The retention index system applied to headspace SPME was recently further studied by Schäfer and coworkers [33]. A linear relationship between logarithm of distribution coefficient and retention index was found for 6 different fiber coatings. This method can, in theory, be applied to any sample form. Furthermore, it has great potential for flavor analysis, although it still needs to be thoroughly validated.

POTENTIAL APPLICATIONS

Eisert and Pawliszyn have reviewed the theory, optimization, and application of SPME in various fields, describing not only its use with GC but also with HPLC. Additionally, several other recent developments have been reported that promise to broaden the range of applicability and increase the effectiveness of SPME in flavor studies [34].

When assisting flavor creation efforts, analytical objectives are generally the identification and quantitation of individual flavor components present in foods, beverages, etc. However, the flavor profile of a food sample depends not only on the total amount of flavor components present, but also on the physical and chemical properties of the sample matrix. For example, a flavor created for a full-fat food generally produces a very different sensory perception among tasters than it does in corresponding low- or no-fat foods. Therefore, to develop a flavor with the desired sensory profile, the flavor release properties of the food matrix need to be studied. Although the strong matrix effects of SPME tend to be a drawback in quantitative analysis, these same effects could be very useful in elucidating flavor/food matrix interactions as part of flavor release studies. The higher sensitivity of SPME for semivolatile components complements conventional headspace analysis. The matrix dependence of SPME analysis can also guide the reverse engineering approach to flavor creation, i.e., modification of an existing flavor for a new or modified food matrix. SPME was applied for analysis of human breath; ethanol, acetone, and isoprene were detected in the concentration range 0.3–5.8 nmol/L [35]. The same approach could also be used to study flavor release *in vivo* during food consumption.

SUMMARY

The general advantages of SPME as a rapid, simple, solventless, and inexpensive method can be readily utilized in food/flavor analysis. Adsorption of analytes on commercially available SPME fibers depends largely on their polarity and molecular size, but also on the physical and chemical properties of the sample matrix. SPME is being increasingly used as a rapid screening method for the analysis of foods, beverages, ingredients, and other botanicals. Although SPME is not recommended for quantitative multicomponent analysis, as involved in flavor duplication work, quantitative analysis can be carried out where necessary (e.g., target analyte analysis) using standard addition or stable isotope dilution procedures. Quantitative analysis using SPME is also possible by employing headspace analysis with the full evaporation technique, and GC retention index-based distribution coefficient estimates. The high selectivity of SPME fibers and strong matrix effects are also potentially useful for flavor release studies.

REFERENCES

1. X Yang and TL Peppard. Solid-Phase Microextraction of flavor compounds - a comparison of two fiber coatings and a discussion of the rules of thumb for adsorption. LC-GC 13: 882-886,1995.
2. X Yang and TL Peppard. Solid-Phase Microextraction for flavor analysis. J Agric Food Chem 42: 1925-1930, 1994.
3. L Pan, M Adams, and J Pawliszyn. Determination of fatty acids using Solid-Phase Microextraction. Anal Chem 67: 4396-4403, 1995.
4. L Pan and J Pawliszyn. Derivatization/Solid-Phase Microextraction: new approach to polar analytes. Anal Chem 69: 196-205, 1997.
5. SB Hawthorne, DJ Miller, J Pawliszyn, and CL Arthur. Solventless determination of caffeine in beverages using Solid-Phase Microextraction with fused-silica fibers. J Chromatogr 603: 185-191, 1992.
6. Y Liu, Y Shen, and ML Lee. Porous layer Solid Phase Microextraction using silica bonded phases. Anal Chem 69: 190-195, 1997.
7. D Louch, S Motlagh, and J Pawliszyn. Dynamics of organic compound extraction from water using liquid-coated fused silica fibers. Anal Chem 64: 1187-1199, 1992.
8. WM Coleman III. A study of the behavior of Mailard reaction products analyzed by Solid-Phase Microextraction-gas chromatography-mass selective detection. J Chromatographic Sci 34: 213-218, 1996.
9. AJ Matich, DD Rowan, and NH Banks. Solid Phase Microextraction for quantitative headspace sampling of apple volatiles. Anal Chem 68: 4114-4118, 1996.
10. JA Field, G Nickerson, DD James, and C Heider. Determination of essential oils in hops by headspace Solid-Phase Microextraction. J Agric Food Chem 44: 1768-1772, 1996.
11. A Steffen and J Pawliszyn. Analysis of flavor volatiles using headspace Solid-Phase Microextraction. J Agric Food Chem 44: 2187-2193, 1996.
12. B Schäfer, P Hennig, and W Engewald. Analysis of monoterpenes from conifer needles using Solid Phase Microextraction. J High Resol Chromatogr 18: 587-592, 1995.
13. E Ibañez and RA Bernhard. Solid-Phase Microextraction (SPME) of pyrazines in model reaction systems. J Sci Food Agric 72: 91-96, 1996.
14. L-K Ng, M Hupé, J Harnois, and D Moccia. Characterization of commercial vodkas by Solid-Phase Microextraction and gas chromatography/mass spectrometry analysis. J Sci Food Agric 70: 380-388, 1996.
15. JS Elmore, MA Erbahadir, and DS Mottram. Comparison of dynamic headspace concentration on Tenax with Solid Phase Microextraction for the analysis of aroma volatiles. J Agric Food Chem 45: 2638-2641, 1997.
16. D Ulrich, S Eunert, E Hoberg, and A Rapp. Analyse des erdbeeraromas mittels festphasen-mikroextraktion. Deutsche Lebensmittel-Rundschau. 91(11): 349-351, 1995.
17. D Picque, A Normand, and G Corrieu. Evaluation of the Solid Phase Microextraction for the direct analysis of aroma compounds in banana. Bioflavour 95, Dijon (France), Ed. INRA, Paris, 117-120, February 14-17, 1995.
18. J Song, BD Gardner, JF Holland, and RM Beaudry. Rapid analysis of voaltile flavor compounds in apple fruit using SPME and GC/time-of-flight mass spectrometry. J Agric Food Chem 45: 1801-1807, 1997.
19. JA Pickett, J Coates, and FR Sharpe. Distortion of essential oil composition during isolation by steam distillation. Chem & Ind 571-572, 1975.

20. WM Coleman III and BM Lawrence. A comparison of selected analytical approaches to the analysis of an essential oil. Flavour & Fragrance J 12: 1-8, 1997.
21. F Pelusio, T Nilsson, L Montanarella, R Tilio, B Larsen, S Facchetti, and JØ Madsen. Headspace Solid-Phase Microextraction analysis of volatile organic sulfur compounds in black and white truffle aroma. J Agric Food Chem 43: 2138-2143, 1995.
22. KG Miller, CF Poole, and TMP Pawlowski. Classification of the botanical origin of cinnamon by Solid-Phase Microextraction and gas chromatography. Chromatographia 42: 639-646, 1996.
23. WM Coleman III. A study of the behavior of polar and nonpolar Solid-Phase Micro-extraction fibers for the extraction of Maillard reaction products. J Chromatographic Sci 35: 245-258, 1997.
24. BD Page and G Lacroix. Application of Solid Phase Microextraction to the headspace gas chromatographic analysis of halogenated volatiles in selected foods. J Chromatogr 648: 199-211, 1993.
25. L Urruty, M Montury, M Braci, J Fournier, and J-M Dournel. Comparison of two recent solventless methods for the determination of procymidone residues in wine: SPME/GC-MS and ELISA tests. J Agric Food Chem 45: 1519-1522, 1997.
26. Y Wang, M Bonilla, and HM McNair. Solid Phase Microextraction associated with microwave assisted extraction of food products. J High Resol Chromatogr 20: 213-216, 1997.
27. JT Clark and JE Bunch. Qualitative and quantitative analysis of flavor additives on tobacco products using SPME-GC-Mass Spectroscopy. J Agric Food Chem 45: 844-849, 1997.
28. C Fischer and U Fischer. Analysis of cork taint in wine and cork material at olfactory subthreshold levels by Solid Phase Microextraction. J Agric Food Chem 45: 1995-1997, 1997.
29. J Ai. Solid Phase Microextraction for quantitative analysis in nonequilibrium situations. Anal Chem 69: 1230-1236, 1997.
30. J Ai. Headspace Solid Phase Microextraction. Dynamics and quantitative analysis before reaching a partition equilibrium. Anal Chem 69: 3260-3266, 1997.
31. M Markelov, and JP Guzowski, Jr. Matrix independent headspace gas chroma-tographic analysis. The full evaporation technique. Anal Chim Acta 276: 235-245, 1993.
32. PA Martos, A Saraullo, and J Pawliszyn. Estimation of air/coating distribution coeffi-cients for Solid Phase Microextraction using retention indexes from linear tempera-ture-programmed capillary gas chromatography. Application to the sampling and analysis of total petroleum hydrocarbons in air. Anal Chem 69: 402-408, 1997.
33. B Schäfer, P Hennig, and W Engewald. Methodological aspects of headspace SPME: application of the retention index system. J High Resol Chromatogr 20: 217-221, 1997.
34. R Eisert and J Pawliszyn. New trends in Solid-Phase Microextraction. Crit Revs Anal Chem 27(2): 103-135, 1997.
35. C Grote and J Pawliszyn. Solid-Phase Microextreaction for the analysis of human breath. Anal Chem 69: 587-596, 1997.

7

Forensic and Toxicology Applications

José R. Almirall and Kenneth G. Furton
Florida International University, Miami, Florida

INTRODUCTION

The typical arsonist is not aware that the application of Solid Phase Microextraction to the analysis of fire debris has lowered the detection limits for accelerants. Crime laboratories are capable of detecting and identifying flammable and combustible liquid residues from the debris of an arson even when the fire consumes most of the accelerant. Typically, these compounds are isolated by passive headspace extraction methods using an activated charcoal strip as the adsorption medium for subsequent elution with a solvent. However, applying SPME as an extraction method has improved the detection of accelerants and results in other advantages compared to current methods.

The analysis of post-blast debris from a bombing scene for the presence of explosive residues is usually conducted with solvent extraction or solid phase extraction methods. Application of SPME for extraction of these compounds has also shown improvements in their detection. Direct sampling methods from aqueous extracts permits relatively clean extractions of both organic and inorganic components from the debris at very low levels.

Concentration of species of toxicological interest from body fluids using direct SPME also provides cleaner extractions of drugs, poisons, and metabolites with reduced sample preparation compared to traditional methods.

FORENSIC APPLICATIONS

SPME has many advantages when applied to forensic specimens. For example, SPME allows multiple sampling and preservation of the sample while minimizing the risk of sample contamination because minimal or no sample handling is required by the technique. SPME can yield faster case turnaround time and is often simpler than traditional techniques; moreover, it is readily automated. Furthermore, the lower detection limits possible using SPME allow confirmation of positive samples that previously went undetected. Finally, the elimination of solvents can save forensic science laboratories money and reduce the risk of analysts being exposed to toxic substances.

The majority of forensic science applications of SPME have been in the area of forensic toxicology described in the next section. Recent forensic SPME applications have included accelerant, explosive, and drug odor analysis. Interestingly, a coated wire adsorption technique was being applied to accelerant analysis in the same year that SPME was being introduced by Pawliszyn [1]. This early technique involved the heated headspace (70° or 80°C) adsorption of accelerants onto carbon-coated aluminum or copper wire followed by n-pentane elution with ultrasonic vibration [2]. Solventless SPME was first applied to arson analysis by the authors in 1994 [3] in which the improved sensitivity for the recovery of light, medium, and heavy petroleum distillates was demonstrated. Compared to the established activated charcoal strip (ACS)/solvent elution method, SPME provided significantly reduced analysis times and the elimination of toxic solvents [4]. SPME analyses of gasoline and kerosene have been compared to headspace, cold trap, and solvent extraction methods and have provided accurate information with less interference peaks [5]. A more detailed study of gasoline, the most commonly used accelerant, confirmed the utility of the SPME technique including lack of interference problems in the presence of wood or plastic pyrolysis products and the ability of SPME to provide reproducible multiple analyses from a single sample [6]. All component peaks required for identification were seen using SPME although, in some cases, their relative ratios were different when compared to ACS. Nevertheless, this should not be a problem in actual casework, as reference chromatograms using SPME should be used for control sample comparison and analysis.

The recovery of accelerants directly from aqueous solvents has also been demonstrated and used in actual casework for lighter fluid, gasoline, and diesel fuel [7]. SPME proved to be more than an order of magnitude more sensitive than the conventional solvent extraction method on 500 ppb solutions; furthermore, it allowed for positive identification of aqueous diesel fuel which was not possible with solvent extraction. SPME has also been used to identify the presence of gasoline in a real arson-suspected fire debris sample, while conventional methods, such as static headspace, lacked adequate sensitivity for the analysis so that accelerants were not detected [8]. Recently, a method for recovering and identifying accelerants from human skin has been developed using 100 μm PDMS fibers with gentle heating for 5 minutes followed by 10 minute sampling from a plastic bag shrouding the suspected hand [9]. The recoverable accelerant was highly dependent on the initial amount, environmental conditions, accelerant type, and time since application. This method may allow for the sampling of a suspect's hand at the scene of a suspicious fire, as seen in Figure 7.1 for a 10 µl gasoline sample on a hand with peak identities in Table 7.1.

Table 7.1 Identity of Hydrocarbons Recovered from Human Skin with SPME

Peak No.	Retention Time (min.)	Compound	Structure
1	3.18	Toluene	
2	4.72	p-xylene	
3	5.25	Nonane	$CH_3(CH_2)_7CH_3$
4	6.90	3-ethytoluene	
5	7.47	2-ethyltoluene	
6	8.07	1,2,4-trimethylbenzene	
7	8.29	Decane	$CH_3(CH_2)_8CH_3$
8	9.18	1,2,3-trimethylbenzene	
9	10.70	Butylbenzene	
10	12.65	Undecane	$CH_3(CH_2)_9CH_3$
11	16.14	Naphthalene	
12	16.88	Dodecane	$CH_3(CH_2)_{10}H_3$
13	20.49	1-methylnaphthalene	
14	21.05	2-methylnaphthalene	

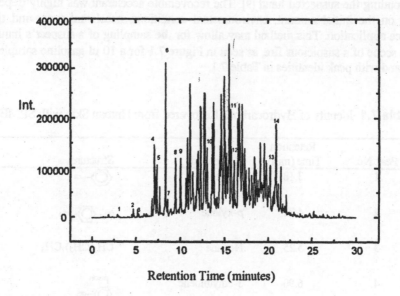

Figure 7.1 SPME/GC/FID of 10 µl gasoline placed on hand (From Ref. 9).

Analysis of semi-volatile compounds, including nitrobenzene and dinitro-toluenes, in water has been reported using a poly(dimethylsiloxane) (PDMS)-coated fiber [10] and GC/FID analysis with detection limits reported as 9-15 ng/mL. Recently, the authors have demonstrated the utility of headspace and direct aqueous immersion SPME for the recovery of various explosives using SPME followed by GC/MS and LC/UV. The direct immersion yielded 1 to 2 orders of magnitude greater sensitivity than headspace [11]. In general, PDMS/DVB (divi-nylbenzene) proved to be the best fiber type for the 14 explosives as shown in Figure 7.2 with results varying depending on the explosive extracted [12]. SPME/GC/ECD generally yielded the best separations and highest sensitivity with detection limits in the low ppb range and demonstrated ability to recover explosive residues from actual post-blast debris as shown in Figure 7.3 for C-4 [13]. One of the major advantages of the direct SPME technique for explosives from acetoni-trile/water solutions is the obvious simple chromatogram obtained, devoid of inter-ferences, from endogenous compounds.

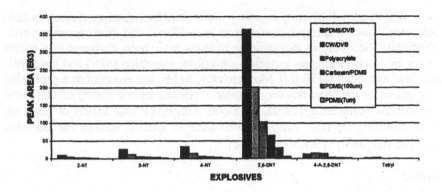

Figure 7.2 Effect of SPME fiber types on the extraction of various explosives (From Ref. 12).

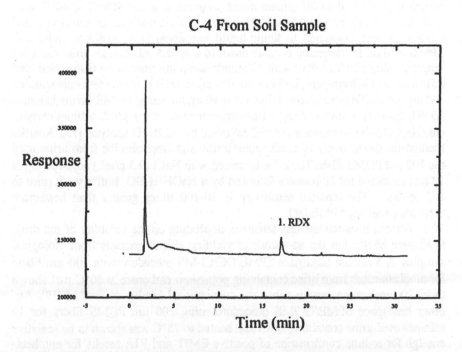

Figure 7.3 SPME Recovery of RDX from C-4 post-blast debris sample (From Ref. 13).

TOXICOLOGY APPLICATIONS

The extraction of many organic poisons from biological specimens, including urine, serum, and blood, continues to be a difficult and time consuming task. Traditional liquid-liquid extraction techniques have been challenged in recent years, by newer techniques including solid-phase extraction (SPE) and supercritical fluid extraction (SFE) [14]. Most recently, SPME has emerged as a promising alternative for the rapid recovery of organic poisons from a variety of biological matrices including urine, plasma, and blood. Although a wide variety of fibers are now available, the most widely applied fiber in forensic science has been the original 100 μm poly(dimethylsiloxane).

Urine

SPME has been applied most extensively to the recovery of drugs from urine. SPME/GC has been successfully applied to the recovery of methadone, benzodiazepines, cannabinoids, phencyclidine, methaqualone, amphetamines, and their metabolites from urine, but it was not as successful for the recovery of cocaine or opiates and their metabolites [15]. The method involved extracting urine samples hydrolyzed at pH 2 or enzymatically digested at pH 4.8, followed by adjusting to pH 12, then 20 minute direct immersion using SPME at 40°C with (PDMS) and poly(acrylate) fibers. This approach yielded greater sensitivity and cleaner extracts compared to liquid-liquid extraction [15]. A direct immersion SPME method for detecting cocaine down to 6 ng/0.5 mL human urine has been reported using a PDMS fiber with 30 minute sampling from urine with added NaF followed by GC/nitrogen-phosphorous detection (NPD) [16]. Methamphetamine and amphetamine were detected down to ~ 10 ng/mL using PDMS/divinylbenzene (DVB) fibers to extract the drugs from urine treated with 1 g/mL sodium carbonate (Na_2CO_3) for 30 minutes at 65°C followed by GC/NPD analysis [17]. Another method for the recovery of methamphetamine and amphetamine from urine used the 100 μm PDMS fiber. The urine is treated with NaCl (0.5 g/mL) adjusted to pH 12 and extracted for 20 minutes followed by a $NaOH-H_3BO_4$ buffer wash prior to GC analysis. The reported sensitivity is 10-100 times greater than headspace methods including SPME [18].

Although sometimes less sensitive, depending on the volatility of the drug, headspace SPME has the advantage of yielding cleaner extracts from biological samples. A 5 minute headspace SPME/GC/CI-MS procedure with 100 μm fibers for amphetamines from urine containing potassium carbonate at 80°C was shown to be 20 times more sensitive than the conventional headspace method [19]. Another headspace SPME/GC/MS procedure using 100 μm PDMS fibers for 15 minutes over urine containing NaCl and heated to 75°C was shown to be sensitive enough for routine confirmation of positive EMIT and RIA results for amphetamine and its analogs (MA, MDMA, MDEA) [20]. Nicotine and cotinine in urine have been analyzed down to 5 and 300 ng/mL, respectively, using 100 μm PDMS headspace SPME/GC/MS and a 5 minute extraction at 80°C with the addition of potassium carbonate [21]. Tricyclic antidepressants in urine have been analyzed down to 24–38 ng/mL using 100 μm PDMS fibers with a 15 minute headspace extraction at 100°C with the addition of sodium hydroxide. GC-FID was used for analysis [22]. Meperidine (pethidine) in urine and blood have been analyzed down to 20 and 100 ng/mL, respectively, using 100 μm PDMS headspace SPME/GC-FID with a 30 minute extraction at 100°C and the addition of sodium hydroxide and sodium chloride [23]. Non routine volatiles including methylene chloride and petroleum products were confirmed in urine and a gastric sample using 100 μm PDMS headspace SPME/GC/MS with a 10 minute extraction at 60°C and the addition of sodium chloride [24]. The absence of an air peak in the GC/MS afforded by the SPME procedure offered a tremendous advantage for identifying the unknown volatiles, which proved to be crucial evidence in the investigation of traffic fatalities.

Blood and Plasma

The free concentration of valproic acid in human plasma has been determined at 1,000 ng/mL with 100 μm PDMS direct immersion SPME/GC-FID using a 3 minute extraction from a sample previously dialyzed for 25 minutes and adjusted to pH 2.5 [25]. Ten antidepressant drugs and metabolites in human plasma were analyzed down to 90–200 ng/mL using 100 μm PDMS headspace SPME/GC-NPD using a 10 minute extraction at 22°C with added sodium hydroxide, followed by 20 second washes with water and 50% methanol prior to GC injection [26]. High protein binding appeared to be the limiting mechanism for better extractions. Methamphetamine and amphetamine in blood were analyzed down to 10 ng/mL using 100 μm PDMS headspace SPME/GC/MS using a 5 minute extraction at 80°C with the addition of sodium hydroxide [27].

Four tricyclic antidepressants in blood have been analyzed down to 61–2,000 ng/mL using 100 μm PDMS headspace SPME/GC-FID and a 60 minute extraction at 100°C with the addition of sodium hydroxide [28]. SPME was successfully applied for detecting ethanol in human body fluids, including urine and blood, down to 10 and 20 μg/mL, respectively. The procedure used 65 μm Carbowax/divinylbenzene headspace SPME/GC-FID with a 15 minute extraction at 70°C and the addition of $(NH_4)_2SO_4$ [29]. An automated headspace SPME method for analyzing blood alcohol concentration showed excellent precision and linearity using 65 μm Carbowax/divinylbenzene headspace SPME/GC-FID with 3 minute exposures. Results were in close agreement with the conventional static headspace method [30].

Unconventional Poisons in Biological Specimens

Although the most common toxicological analysis involves drug extraction and recovery, drugs actually represent only about 60% of the estimated 1,600 possible poisons [14]. The second largest class of poisons are the pesticides representing about 30% of possible compounds followed by anions, metals, gases, and volatiles, which comprise a total of less than 10% of possible poisons. The relatively limited number of compounds in this last group have resulted in well-established screening and detection methods; however, the many advantages of SPME may lead to improved analysis methods for many of these poisons in the near future. The more than 1,500 possible drugs and pesticides make the development of routine screening and detection methods especially challenging. Approximately 20% of possible drugs are detected in forensic toxicology laboratories while only 4% of possible pesticides are detected. This discrepancy may be indicative of the lack of routine methods for pesticide analysis performed in many forensic science labs.

Even though few specific SPME methods for pesticide analysis from biological specimens have been reported, numerous environmental applications have been developed that may be modified for toxicological use. Recent examples include methods with gas chromatography followed by flame ionization detection [31], electron capture detection [32], atomic emission detection [33], thermionic detection [34], nitrogen-phosphorous detection [35,36,37], mass spectrometry [37,38,39,40], and finally, a validation of pesticide analysis by SPME/GC/MS via a round robin study [41]. Organophosphate pesticides in blood and urine have been analyzed down to 5-80 ng/mL and 2-24 ng/mL, respectively, using 100 μm PDMS headspace SPME/GC/NPD and a 20 minute extraction at 100°C with the addition of HCl only for blood, and HCl, NaCl, and $(NH_4)_2SO_4$ for urine [42]. The common organophosphorous pesticide, malathion, has been detected in blood down to 1,000 ng/mL using 100 μm PDMS headspace SPME/GC/MS and a 5 minute extraction at 90°C with the addition of $(NH_4)_2SO_4$ and H_2SO_4 [43]. Six carbamate pesticides in blood and urine were analyzed down to 100-500 ng/mL and 10-50 ng/mL, respectively, using 100 μm PDMS headspace SPME/GC/NPD and a 30 minute extraction at 70°C with the addition of NaCl [44].

A headspace SPME/GC/MS method has been developed for the analysis of toluene, xylenes, and C_9-C_{15} hydrocarbons down to 100 ng/mL and C_{16}-C_{20} hydrocarbons down to 1,000 ng/mL in human blood [45]. The method involved heating the blood samples with 10% NaOH to 90°C followed by cooling to room temperature for adsorption of both low and high molecular weight analytes. Moreover, it was successfully applied to the medico-legal autopsy of a fire victim with kerosene substances detected. The analysis of thinner components in whole blood and urine down to 1 ng/mL has been performed using 100 μm PDMS headspace SPME/GC-FID with a 5 minute extraction at 80°C [46]. Ten local anesthetics have been analyzed down to 58-830 ng/mL using 100 μm PDMS headspace SPME/GC-FID with a 40 minute extraction at 100°C from human blood deproteinized with perchloric acid and sodium hydroxide and $(NH_4)_2SO_4$ added [47]. The SPME/GC analysis of the highly toxic methylmercury in aqueous and tissue samples has been reported using MS and atomic fluorescence detection [48, 49].

MISCELLANEOUS APPLICATIONS AND FUTURE DEVELOPMENTS

A modified SPME device is capable of detecting down to 5.8 nmol/L ethanol, 1.8 nmol/L acetone, and 0.3 nmol/L isoprene in human breath using a 65 μm PDMS/DVB fiber and 1 minute sampling followed by GC/MS analysis [50]. Moreover, SPME is a useful tool to determine phospholipid/water partition coefficients and free (bioavailable) concentrations *in vitro* systems, which may make experimental data more meaningful for quantitative *in vivo* extrapolations [51]. Chapter 4 describes how SPME has been used to determine residual organic solvents in pharmaceutical samples [52]; furthermore, it has been used to characterize street narcotic odors [53]. Figure 7.4 illustrates the effect of SPME fiber type on the room-temperature recoveries of cocaine odor chemical with recoveries increasing with time up to about 12 hours, and the optimum fiber found to be Carbowax/DVB [54]. SPME has played a vital role in confirming the chemicals responsible for drug dog alerts to suspected drug money with a typical SPME/GC odor signature profile for street cocaine shown in Figure 7.5 [55]. Although most U.S. currency in circulation is contaminated with μg quantities of cocaine [56], these levels appear to be well below the threshold levels of drug detector dogs.

Comparison of GC/FID Response of Cocaine Odors Extracted by Various Fibers for 12 Hours

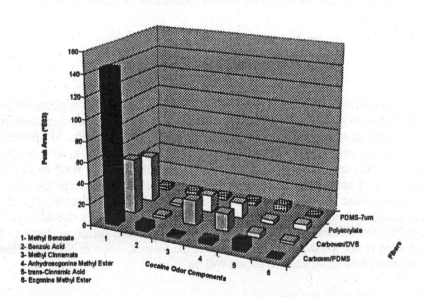

Figure 7.4 Effect of SPME fiber types on the extraction of cocaine odors (From Ref. 54).

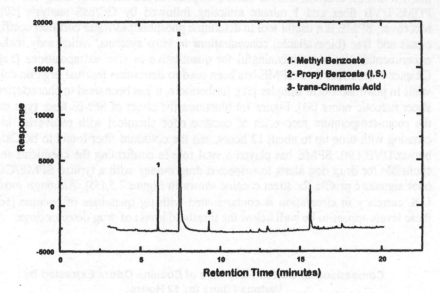

Figure 7.5 SPME/GC recovery of street cocaine odor chemicals (From Ref. 55).

Future developments of SPME for forensic and toxicology applications will likely include further improvements in automation and field sampling combined with rapid on-site analysis including Fast GC and HPLC methods. SPME/HPLC methods for most analytes including poisons and explosives are, to date, undeveloped. Field portable SPME/GC and SPME/LC methods could revolutionize the anti-terrorist battle and dramatically improve law enforcement response time by providing rapid on-site confirmation of drugs, accelerants, explosives, etc. Field air sampling combined with portable chromatographic analysis should have far reaching implications from chemical weapons to disease detection. Indeed, the question is not whether SPME will impact the forensic sciences and toxicology, but rather, how significant the impact will be.

REFERENCES

1. CL Arthur and J Pawliszyn. Solid phase microextraction with thermal desorption using fused silica optical fibers. Anal Chem 62:2145-2148, 1990.
2. DJ Tranthim-Fryer. The application of a simple and inexpensive modified carbon wire adsorption/solvent extraction technique to the analysis of accelerants and volatile organic compounds in arson debris. J Forensic Sci 35:271-280, 1990.
3. JR Almirall, KG Furton, and JC Bruna. A novel method for the analysis of gasoline from fire debris using headspace Solid-Phase microextraction, Southern Association of Forensic Scientists - Fall 1994 Meeting, Orlando, Florida, September 7 - 10, 1994.
4. KG Furton, JR Almirall, and J Bruna. A simple, inexpensive, rapid, sensitive and solventless method for the recovery of accelerants from fire debris based on SPME. J High Resolut Chromatogr 18:625-629, 1995.
5. T Kaneko and M Nakada. Forensic application of the solid phase microextraction method to the analysis of gasoline and kerosene. In: Reports of the National Research Institute of Police Science: Research on Forensic Science. 48:107-111, 1995 (Japanese).
6. KG Furton, JR Almirall, and J Bruna. A novel method for the analysis of gasoline from fire debris using headspace solid-phase microextraction. J Forensic Sci 41:12-22, 1996.
7. JR Almirall, J Bruna, and KG Furton. The recovery of accelerants in aqueous samples from fire debris using solid phase microextraction (SPME). Science and Justice 36:283-287, 1996.
8. A Steffen and J Pawliszyn. Determination of liquid accelerants in arson suspected fire debris using headspace solid-phase microextraction. Anal Communications 33:129-131, 1996.
9. J Wang. Optimization of SPME/GC for the recovery of accelerants from fire debris and human skin. M.S. thesis, Florida International University, Miami, FL, 1998.
10. J-Y Horng and S-D Huang. Determination of the semi-volatile compounds nitrobenzene, isophorone, 2,4-dinitrotoluene and 2,6-dinitrotoluene in water using solid-phase microextraction with a poly(dimethylsiloxane)-coated fiber. J Chromatogr 678:313-318, 1994.
11. M Bi, JR Almirall, and KG Furton. Potential and problems in the analysis of explosives by solid-phase microextraction followed by HPLC and GC/MS. 1997 Annual Meeting of the Florida Sections - American Chemical Society, Orlando, Florida, May 2 - 3 1997.
12. JR Almirall, G Bi, and KG Furton. The analysis of high explosive residues by solid-phase microextraction followed by HPLC, GC/ECD and GC/MS. Proceedings of the 50th Annual Meeting of the American Academy of Forensic Sciences, San Francisco, California, February 9 - 14, 1998, B71.
13. M Bi. The potential of solid phase microextraction (SPME) for the recovery of explosives followed by gas and liquid chromatography. M.S. thesis, Florida International University, Miami, FL, 1998.
14. KG Furton and J Rein. Trends in techniques for the extraction of drugs and pesticides from biological specimens prior to chromatographic separation and detection. Anal. Chimica Acta 236:99-114, 1990.
15. K Singer, B Wenz, V Seefeld, and U Speer. Determination of drugs of abuse using solid-phase microextraction. Labor-Med. (German) 18(2):112-118, 1995.

16. T Kumazawa, K Watanabe, K Sato, H Seno, A Ishii, and O Suzuki. Detection of cocaine in human urine by solid-phase microextraction and capillary gas chromatography
with nitrogen-phosphorus detection. Jpn J Forensic Toxicol 13(3):207-210, 1995.

17. A Ishii, H Seno, T Kumazawa, M Nishikawa, K Watanabe, H Hattori, and O Suzuki.
Simple clean-up of methamphetamine and amphetamine in human urine by direct-
immersion solid phase microextraction (DI-SPME). Jpn J Forensic Toxicol
14(3):2228-232, 1996.

18. K Ameno, C Fuke, S Ameno, H Kinoshita, and I Ijiri. Application of a solid-phase
microextraction technique for the detection of urinary methamphetamine and amphetamine by gas chromatography. Can Soc Forens Sci J 29(2):43-48, 1996.

19. M Yashiki, T Kojima, T Miyazaki, N Nagasawa, Y Iwasaki, and K Hara. Detection of
amphetamines in urine using headspace-solid phase microextraction and chemical
ionization selected ion monitoring. Forens Sci International 76:169-177, 1995.

20. F Centini, A Masti, and IB Comparini. Quantitative and qualitative analysis of
MDMA, MDEA, MA and amphetamine in urine by headspace/solid phase microextraction (SPME) and GC/MS. Forens Sci International 83:161-166, 1996.

21. M Yashiki, N Nagasawa, T Kojima, T Miyazaki, and Y Iwasaki. Rapid analysis of
nicotine and cotinine in urine using headspace-solid phase microextraction and selected ion monitoring. Jpn J Forensic Toxicol 13(1):1724, 1995.

22. T Kumazawa, X-P Lee, M-C Tsai, H Seno, A Ishi, and K Sato. Simple extraction of
tricyclic antidepressants in human urine by headspace solid-phase microextraction
(SPME). Jpn J Forensic Toxicol 13(1):25-300, 1995.

23. H Seno, T Kumazawa, A Ishii, M Nishikawa, H Hattori, and O Suzuki. Detection of
meperidine (pethidine) in human blood and urine by headspace solid phase microextraction and gas chromatography. Jpn J Forensic Toxicol 13(3):211-215, 1995.

24. WE Brewer, RC Galipo, SL Morgan, and KH Habben. The confirmation of volatiles
by solid-phase microextraction and GC-MS in the investigation of two traffic fatalities.
J Anal Tox 21:286-290, 1997.

25. M Krogh, K Johansen, F Tonnesen, and KE Rasmussen. Solid-phase microextraction
for the determination of the free concentration of valproic acid in human plasma by
capillary gas chromatography. J Chromatogr B 673:299-305, 1995.

26. S Ulrich and J Martens. Solid-phase microextraction with capillary gas-liquid chromatography and nitrogen-phosphorus selective detection for the assay of antidepressant drugs in human plasma. J Chromatogr B 696:217-234, 1997.

27. N Nagasawa, M Yashiki, Y Iwasaki, K Hara, and T Kojima. Rapid analysis of amphetamines in blood using headspace-solid phase microextraction and selected ion
monitoring. Forens Sci International 78:95-102, 1996.

28. X-P Lee, T Kumazawa, K Sato, and O Suzuki. Detection of tricyclic antidepressants in
whole blood by headspace solid-phase microextraction and capillary gas chromatography. J Chromatogr Sci 35(7):302-308, 1997.

29. T Kumazawa, H Seno, X-P Lee, A Ishii, O Suzuki, and K Sato. Detection of ethanol in
human body fluids by headspace solid-phase microextraction (SPME)/capillary gas
chromatography. Chromatographia 43 (7/8):393-397, 1996.

30. Z Penton. Blood alcohol determination with automated solid phase microextraction
(SPME): a comparison with static headspace sampling. Can Soc Forens Sci J 30(1):7-
12, 1997.

31. MT Sng, FK Lee, and HA Lakso. Solid-phase microextraction of organophosphorous
pesticides from water. J Chromatogr A 759:225-230, 1997.

32. S Magdic and JB Pawliszyn. Analysis of organochlorine pesticides using solid-phase microextraction. J Chromatogr A 723:111-122, 1996.

33. R Eisert, K Levsen, and G Wunsch. Element-selective detection of pesticides by gas chromatography-atomic emission detection and solid-phase microextraction. J Chromatogr A 683:175-183, 1994.

34. V Lopez-Avila, R Young, and WF Beckert. On-line determination of organophosphorous pesticides in water by solid-phase microextraction and gas chromatography with thermionic detection. J High Resol Chromatogr 20:487-492, 1997.

35. R Eisert and K Levsen. Determination of organophosphorous, triazine and 2,6-dinitroaniline pesticides in aqueous samples via solid-phase microextraction (SMPE) and gas chromatography with nitrogen-phosphorous detection. Fresenius J Anal Chem 351:555-562, 1995.

36. I Valor, JC Molto, D Apraiz, and G Font. Matrix effects on solid-phase microextraction of organophosphorous pesticides from water. J Chromatogr A 767:195-203, 1997.

37. TK Choudhury, KO Gerhardt, and TP Mawhinney. Solid-phase microextraction of nitrogen- and phosphorous-containing pesticides from water and gas chromatographic analysis. Environ Sci Technol 30:3259-3265, 1996.

38. R Eisert and K Levsen. Determination of pesticides in aqueous samples by solid-phase microextraction in-line coupled to gas-chromatrography-mass spectrometry. J Am Soc Mass Spectrom 6:1119-1130, 1995

39. L Urruty and M Montury. Influence of ethanol on pesticide extraction in aqueous solutions by solid-phase microextraction. J Agric Food Chem 44:3871-3877, 1996.

40. AA Boyd-Boland, S Magdic, and JB Pawliszyn. Simultaneous determination of 60 pesticides in water using solid-phase microextraction and gas chromatography-mass spectrometry. Analyst 121:929-938, 1996.

41. T Gorecki, R Mindrup, and J Pawliszyn. Pesticides by solid-phase microextraction: Results of a round robin test. Analyst 121:1381-1386, 1996.

42. X-P Lee, T Kumazawa, K Sato, and O Suzuki. Detection of organophosphorous pesticides in human body fluids by solid-phase microextraction (SPME) and capillary gas chromatography with nitrogen-phosphorus detection. Chromatographia 42(3/4):135-140, 1996

43. A Namera, M Yashiki, N Nagasawa, Y Iwasaki, and T Kojima. Rapid analysis of malathion in blood using headspace solid-phase microextraction and selected ion monitoring. Forensic Sci International 88:125-131, 1997.

44. H Seno, T Kumazawa, A Ishii, M Nishika, K Watanabe, H Hattori, and O Suzuki. Determination of some carbamate pesticides in human body fluids by headspace solid phase microextraction and gas chromatography. Jpn J Forensic Toxicol 14(3):199-203, 1996.

45. Y Iwasaki, M Yashiki, N Nagasawa, T Miyazaki, and T Kojima. Analysis of inflammable substances in blood using headspace-solid phase microextraction and chemical ionization ion monitoring. Jpn J Forensic Toxicol 13(3):189-194, 1995.

46. X-P Lee, T Kumazawa, and K Sato. A simple analysis of 5 thinner components in human body fluids by headspace solid-phase microextraction (SPME). Int J Legal Med 107:310-313, 1995.

47. T Kumazawa, X-P Lee, K Sato, H Seno, A Ishii, and O Suzuki. Detection of ten local anaesthetics in human blood using solid-phase microextraction (SPME) and capillary gas chromatography. Jpn J Forensic Toxicol 13(3):182-188, 1995.

48. Y Cai and JM Bayona. Determination of methylmercury in fish and river samples using in situ sodium tetraethylborate derivatization followed by solid-phase microextraction and gas chromatography-mass spectrometry. J Chromatogr A 696:113-122, 1995.

49. Y Cai, S Monsalud, KG Furton, R Jaffe, and RD Jones. Determination of methylmercury in fish and aqueous samples using solid-phase microextraction followed by gas chromatography-atomic fluorescence spectrometry. Applied Organometallic Chemistry 12:565-569, 1998.

50. C Grote and J Pawliszyn. Solid-phase microextraction for the analysis of human breath. Anal Chem 69:587-596, 1997.

51. WHJ Vaes, EU Ramos, C Hamwijk, IV Holsteijn, BJ Blaauboer, W Seinen, HJM Verhaar, and JLM Hermens. Solid phase microextraction as a tool to determine membrane/water partition coefficients and bioavailable concentrations in vitro systems. Chem Res Toxicol 10:1067-1072, 1997.

52. RF Mindrup. Solid phase microextraction simplifies preparation of forensic, pharmaceutical, and food and beverage samples. Chem New Zealand 3:21-23, 1995.

53. KG Furton, Y-L Hsu, T Luo, N Alvarez, and P Lagos. Novel sample preparation methods and field testing procedures used to determine the chemical basis of cocaine detection by canines. In Forensic Evidence Analysis and Crime Scene Investigation. John Hicks, Peter De Forest, Vivian M. Baylor, editors. Proc. SPIE Vol. 2941, 1997, pp. 56-62.

54. KG Furton, Y-L Hsu, T Luo, F Lopez, and S Rose. Diffusion studies and SPME/GC/MS/MS analysis of volatile drug components and the relevance to detector dog alerts to suspected drug money. Proceedings of the 50th Annual Meeting of the American Academy of Forensic Sciences, San Francisco, California, February 9 - 14, 1998, B49.

55. T Luo. SPME/GC signature analysis of cocaine impurities and correlation to drug dog detection. M.S. thesis, Florida International University, Miami, FL, 1998.

56. A Negrusz, JL Perry, and CM Moore, Detection of cocaine on various denominations of United States currency. J Forensic Sci 43(3):626-629, 1998.

8

New Developments in SPME

Ralf Eisert and Janusz Pawliszyn
University of Waterloo, Waterloo, Ontario, Canada

INTRODUCTION

This chapter summarizes new areas where SPME methods are successfully established. Several new devices based on SPME have been developed for air monitoring, integrated sampling, fast gas chromatography, automated SPME, interfacing to HPLC, automated in-tube SPME/HPLC, calibration techniques based on physico-chemical properties of the target compounds, speciation of inorganic analytes, SPME-LC/MS, field sampling devices, on-site analysis, and automated on-line systems. Most of these applications are based on systems that are not yet commercially available for routine labs.

Solid-phase microextraction is not restricted to screening purposes. In fact, different calibration methods are leading to a high precision in quantitative analysis. The fast transfer of the sample from extraction to separation and detection indicate its high potential for applications where the storage time of classical extraction and sample preparation techniques failed to provide information about biodegradation and very labile intermediates. Real-time analysis, a new area that is an important aspect for flavor and perfume analysis, opens a wide application field for SPME methods. Furthermore, the simple handling of the entire solvent-free sample preparation techniques should be considered a main advantage when on-site and field analysis is under investigation. SPME provides unique characteristics when sampling and determining target compounds in remote areas and/or when instant decisions (accidents) or proof (forensic) data are recommended.

To date, solid-phase microextraction (SPME) has been applied for many different areas in analytical chemistry [1]. The basic concept of SPME was first described by Belardi and Pawliszyn [2], and mathematical models were developed to describe the mass transfer onto the fiber and diffusion phenomena that determine the kinetics of extraction [1,5]. The technique was made practical by placing the fiber in a microsyringe [3] to create an SPME device, which was commercialized by Supelco in 1993. Early applications included air, water, and soil analyses [6,11,12]; accordingly, methods were developed exclusively by coupling SPME with GC [3-9]. In this same year, automation of the entire SPME/GC system was achieved by modifying a commercial GC autosampler to perform direct sampling [4,13]. This system could perform not only the sample preparation, but also the entire analysis, hence significantly increasing laboratory throughput. In 1995, the first interface for SPME/HPLC coupling was described [10].

Presently, SPME applications utilizing either GC or LC are still primarily based on the manual sampling device. This is due to insufficient temperature control and a lack of agitation techniques that slowed the initial rapid growth of automated SPME/GC applications. Nevertheless, within the past two years many new devices and coupling techniques have been developed to overcome initial problems (e.g., agitation for the automated SPME device [14], field portable devices [15], and analysis of thermally labile organic compounds by successful, automated hyphenation to HPLC [16] and inorganic target analytes [17-19]).

Direct Extraction

A typical exposure of the fiber when sampling semi- and non-volatile compounds is direct extraction from a homogeneous aqueous phase. In theory, no headspace is present, but in practice, a small headspace volume always exists in the sampling vial. Nevertheless, it shows no significant deterioration of the analytes in the aqueous phase. Under perfect agitation conditions (e.g., magnetic stirring) the aqueous phase moves very rapidly with respect to the fiber. Theoretically, all analyte molecules present in the sample should have easy access to the fiber coating. The time to reach equilibrium is significantly lower compared to the same experiment done for static (no agitation) conditions. Figure 8.1 shows a typical result obtained when both exposure techniques are compared.

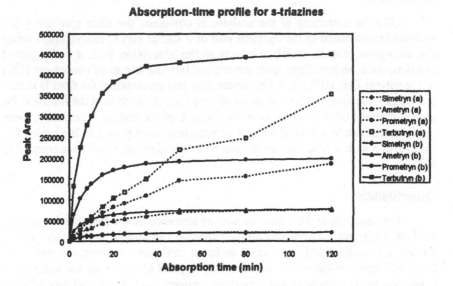

Figure 8.1 Absorption-time profiles for four s-triazines using (a) static absorption conditions and (b) fiber vibration method (for 2 mL vials). The equilibrium is reached substantially faster for agitation techniques (From Ref. 40).

In this example, the equilibration time for compounds with higher fiber coating/sample distribution constant values, K_{fs}, such as prometryn, is reduced from more than 2 hours to 35 minutes. From the theory [5], the time required to reach equilibrium is infinitely long. However, in practice, a change in mass extracted cannot be determined if it is smaller than the experimental error, which is typically 5%. Thus, the equilibration time is assumed to be achieved when 95% of the equilibrium amount of the analyte is extracted from the sample.

Independent from the agitation level and method (magnetic stirring, fiber vibration, flow-through cell, etc.), fluid contacting a fiber's surface is always stationary, and as the distance from the fiber surface increases, the fluid movement gradually increases until it corresponds to the bulk flow of the sample. This model describes the mass flow using this boundary layer concept [5] assuming that, in a defined zone surrounding the fiber, no convection occurs and perfect agitation is everywhere in the bulk solution. The static layer is called a Prandtl boundary layer [20]; moreover, its thickness depends on the agitation speed and viscosity of the fluid. Agitation methods are very successful to reduce the thickness of the layer, which determines the mass transfer into the fiber coating. Thus, the equilibration time is significantly decreased.

Desorption

After the extraction of the analytes is complete, the fiber containing the analytes is transferred to the injection port of a GC or HPLC instrument. During the desorption process, which is inverse to the absorption from a well-agitated solution, the analyte diffuses from the coating into the stream of carrier gas (GC) or the solvent fluid (HPLC). A high linear flow rate surrounding the fiber is necessary to provide an initial concentration of zero for each analyte in the fluid and for instant removal of the analytes from the vicinity of the coating. Practically, these requirements can be achieved inside a very narrow insert placed in the heated GC injector or a narrow tube used for the injection chamber when SPME is coupled to HPLC.

Quantitation

The equations to determine the amount absorbed by the fiber coating are described in Chapter 1. Depending on the exposure and sample matrix, these calculations can be simplified. In practice, different calibration methods are applicable to SPME. External calibration can be easily applied to determine the initial concentration for homogenous and very clean samples, such as air and groundwater [21,22], when using equilibrium conditions and knowing the compounds' K_{fs} values. For complex matrices or changing matrix composition, standard addition techniques–which compensate for the matrix effect–or isotopically labeled standards should be used to yield higher precision and more reliable data [23]. The system can be calibrated using spiked (aqueous) samples for non-equilibrium conditions and unknown K_{fs} values of the compounds. However, matrix effects are not compensated in this method that is limited to very clean samples. In all cases, the fiber blank will verify that neither the instrument nor the SPME apparatus is contaminated by analytes or interfering compounds.

In addition to experiment, the external calibration can be performed using physico-chemical parameters or chromatographic results. The concept of retention indexes from linear temperature-programmed GC can be applied to estimate air/coating distribution coefficients and calibrate the SPME method based on a single injection [24,25]. In this approach, the column phase is identical to the SPME coating [24]. It was successfully applied to the analysis of organic compounds in air [24] and water [26]. The calibration technique is based on a simple concept which is used to estimate distribution coefficients (K_{fa}) between the matrix (air) and the poly(dimethylsiloxane) SPME fiber coating. The technique uses the linear temperature-programmed retention index system (LTPRI). The linear relationship ($r^2=0.99989$) between the log K for a series of compounds (n-alkanes) and LTPRI is described [24] in equation 8-1:

$$\log K = a + b(LTPRI) \qquad 8.1$$

where a is the y-intercept and b is the slope of the calibration curve (detector response).

Thus, the K value of an unknown compound can be established in a single GC run. Determining the LTPRI of this compound leads automatically to its K value. The concept was developed for 29 isoparaffinic compounds and a group of 33 aromatic compounds; moreover, it was applied to determine a complex mixture of gasoline [24]. The results obtained were compared to standard procedures and showed identical results.

OPERATING PRINCIPLES

Extraction Modes

The selection of extraction mode should be based on the sample matrix, analyte volatility, and its affinity to the matrix. In general, three different exposure techniques exist: headspace or air, direct sampling from the aqueous phase, and direct exposure using membrane protected extraction [27]. Direct air sampling or headspace sampling can be considered for very volatile compounds (VOCs), such as BTEX or purgeables. It is preferred due to faster equilibration times; furthermore, the selectivity is higher when dirty samples are analyzed. Clean, aqueous samples, such as groundwater, can be extracted in the direct extraction mode, especially if semi- and non- volatile compounds will be extracted. For very dirty samples, the fiber can be protected using a membrane protection. In aqueous samples, very polar compounds, such as strong acids and bases, are very difficult to extract. In this case, the adjustment of the pH value and addition of salt can be considered to extract these compounds that have a high affinity toward the matrix, as discussed in Chapter 2.

The diffusion process in air is very fast and only limited by the diffusion in the coating compared to the diffusion in water. Therefore, many volatile analytes reach equilibrium within 5 minutes when sampling from air or headspace. In contrast, the equilibrium for semi-volatile compounds using direct exposure of the fiber to the aqueous sample takes up to 2 hours under static absorption conditions. The equilibration time is significantly reduced when agitation is applied during the extraction process [14,28]. The effectiveness of the agitation process determines the equilibration time in aqueous samples. In non-automated sampling, magnetic stirring is the most commonly used agitation technique. However, one has to ensure that the rotational speed of the stir bar is constant and the base plate is thermally isolated from the vial containing the sample. Otherwise, heating of the sample might result in loss of precision.

Headspace sampling is very fast, and magnetic stirring does not affect the diffusion from the headspace to the fiber. However, when the concentration in the headspace is significantly reduced by the SPME fiber exposure, the mass transport between the aqueous sample and the headspace slows down the extraction process. Agitation facilitates the equilibrium between the headspace and the aqueous phase during the SPME sampling, thus reducing the depletion of the headspace concentration.

The alternatives that are discussed in this chapter should be considered especially for automated or on-line systems, such as fiber vibration and flow-through cell design facilitating a high linear flow at the fiber surface [14]. Varian has implemented the fiber vibration technique in its SPME autosampler for GC.

Desorption Interfaces

Two desorption techniques are used for SPME to transfer the absorbed amount from the fiber coating to a chromatographic column. First, for coupling to gas chromatography, the fiber is exposed to the heated injection port of a GC and then thermally desorbed to release the analytes from the fiber. Second, when coupled to HPLC, the fiber is placed in a small desorption chamber, which is a piece of HPLC tubing, where desorption into an organic solvent releases the compounds from the coating [10,29].

Thermal Desorption in GC

A narrow bore insert is required for fast desorption using a splitless injector or a Septum-Equipped, Programmable Injector (SPI). In addition, a hot, on-column injection can be used. The narrow bore insert sustains a high linear flow around the fiber during the desorption, thus reducing the desorption time. The highest possible desorption temperature that is amenable for the target analytes and the fiber coating should be used for a fast transfer of the analytes. The injection band can be sharpened by using a thick film column, cryofocusing, or a retention gap. Cryotrapping is necessary for a limited number of applications. In general, a 1 μm thick column is sufficient for sharpening the injection band width of volatile organic compounds. The fiber should be exposed immediately after its introduction to the insert and placed at the heated part of the injector (depth).

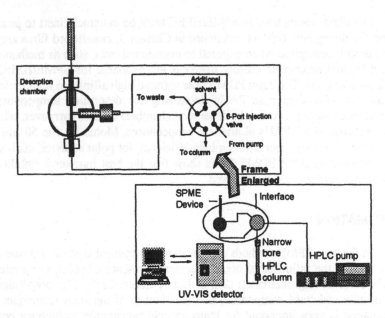

Figure 8.2 Instrumental setup of the manual SPME/HPLC interface. (From Ref. 10.) Reprinted with permission from Analytical Chemistry. Copyright 1995 American Chemical Society.

Solvent Desorption in HPLC

A commercial interface is available from Supelco for manual SPME/HPLC, which is based on the initial loop type injection design [10] used for coupling SPME with HPLC (Figure 8.2). The appropriate solvent selection and flow determine the desorption process. The manual SPME/HPLC interface uses the initial mobile phase composition for the desorption. The linear flow rate should be maximized by choosing a small ID tubing, which is imperative due to the low volumetric flow rate in HPLC. The main disadvantage of this injection concept is carryover, which is mostly related to insignificant desorption conditions. Moreover, the elution power of the initial eluent composition is not sufficient for a quantitative desorption of the absorbed analytes (e.g., high molecular weight compounds (PAHs)). Heating the interface might help to increase the driving force of desorption, but this is limited to the thermal stability of the target compounds. Another approach is to fill the desorption chamber with a pure organic solvent to increase the desorption power. However, peak broadening might be increased, as well. To overcome these effects, a complete separation of the sample desorption step from the transfer to the HPLC column can be performed. This approach is used in the later discussion on the automated in-tube SPME system.

The fiber coating used in SPME/HPLC must be extremely inert to protect it during the desorption step. As mentioned in Chapter 3, crosslinked films are very stable during desorption when exposed to organic solvents, such as methanol. To date, a limited number of coating materials are available for SPME/HPLC that fulfill those criteria. The 7 μm PDMS fiber shows a high affinity to high molecular weight compounds, such as PAHs. Increasing the desorption temperature (by wrapping heating tape around the desorption chamber) reduces carryover, which is often determined for PAHs at ambient temperatures. Moreover, the 50 μm Carbowax template resin fiber shows a high efficiency for polar analytes, such as polar carbamates, and the PDMS fibers show that the best long-term stability and sufficient ruggedness can be achieved.

AUTOMATION

In the past, SPME methods optimization has concentrated on: (1) matrix effects, such as the influence of pH, salt content, presence of polar, low molecular weight solvents (e.g. methanol) [22], or humic material [30]; (2) competition between main and trace compounds; and (3) the use of agitation techniques [14]. Automation is very important for many sample preparation techniques because those that cannot be automated are less often used for routine analysis, even if they offer other attractive features, like high selectivity or sensitivity. However, having no suitable automation of the entire SPME method, especially for SPME-HPLC, is the most often cited disadvantage of SPME. To date, the applications of SPME-HPLC are all based on the manual device and interface. For SPME-GC, automated sample preparation is available using Varian Associates' GC autosampler. It will perform direct sampling (i.e., the fiber is immersed in the liquid sample either by static absorption or by vibrating the fiber in the sample), and headspace sampling, (i.e., the fiber is placed above the sample). The autosampler will accommodate both 2 mL and 16 mL sample vials.

SPME-GC

The automated fiber vibration method works very efficiently, especially when using the small 2 mL vials. Results in Table 8-1 show equilibrium was achieved for all five compounds within 35 minutes when using the 2 mL vials. Even the 16 mL vials show an efficient extraction, but more than 80 minutes was necessary for all compounds to achieve equilibrium. We have to keep in mind that in these experiments no operator intervention was necessary, which shows the great potential of this technique for routine analyses. For example, the analysis of semi-volatiles, such as pesticides, seems to be a promising field for the automated SPME-GC technique.

Table 8.1 Equilibration Times for All Investigated Compounds Obtained Under Different Absorption Modes.

Compound	K_{OW}	Equilibration time [min]						
		2 mL sta[a]	2 mL stir[a]	2 mL vib[a]	16 mL sta[b]	16 mL stir[b]	16 mL vib[b]	40 mL flow[c]
Simetryn	347	60	30	30	80	25	40	30
Ametryn	955	>120	35	30	>120	30	50	30
Prometryn	3236	>120	30	35	>120	30	80	35
Terbutryn	5495	>120	30	35	>120	50	80	35
Parathion	6761	>120	30	20	>120	35	35	25

[a] The results were obtained for 2 mL vials with 1.4 mL sample volume.
[b] The results were obtained for 16 mL vials with 12 mL sample volume.
[c] The results were obtained for 40 mL vials with 10 mL sample volume using the flow-through extraction cell.
Abbreviations: flow = SPME fiber is exposed to the flow-through extraction cell, sta = static absorption with no agitation of the aqueous sample, stir = mixing of the aqueous sample using magnetic stirring, and vib = fiber vibration method using autosampler agitation.
From Ref. 14. Reprinted from Journal of Chromatography A, 1997 with kind permission of Elsevier Science – NL, Sara Burgerhartstraat 25, 1055 KV Amsterdam, The Netherlands.

In general, the results obtained for the 2 mL and 16 mL vials show a similar trend when using different extraction conditions, except for the total amount extracted [14]. Sonication is mentioned in addition to the other extraction processes used in this study. As expected and shown in the past [28], sonication is a very efficient sample agitation technique for SPME. For example, very volatile organic compounds (VOCs), such as toluene, equilibrate in less than 1 minute, which is very close to the theoretical prediction. The major disadvantage of sonication is sample heating during the extraction; consequently, analyte loss is caused by sonication-induced decomposition. To date, sonication is not available for automated SPME; furthermore, the number of applications using sonication is negligible.

The precision results are summarized in Table 8.2 for five repetitive injections when using the 2 mL and 16 mL vials. During the 30 minute absorption time, most of the investigated compounds reach their equilibrium under agitation conditions. The precision obtained for the 2 mL vials when using the fiber vibration method can be considered optimal. The RSDs were < 3% for all compounds when agitation was used. The precision obtained for static absorption conditions is as poor as observed for the 2 mL vials. For semi-volatile compounds, such as s-triazines, the precision is increased with magnetic stirring compared to static absorption.

Table 8.2 Precision Achieved with Autosampler SPME-GC/FID Method of Five Selected Pesticides Using 2 mL and 16 mL Vials and Three Different Absorption Modes.

	Precision (RSD [%])					
Compound	Static[a]	Magnetic Stirring[a]	Fiber Vibration[a]	Static[b]	Magnetic Stirring[b]	Fiber Vibration[b]
Simetryn	0.9	2.6	1.4	1.2	4.2	3.2
Ametryn	3.0	1.3	1.1	2.6	0.7	3.4
Prometryn	7.0	1.1	1.1	4.1	0.7	3.9
Terbutryn	7.8	2.9	1.0	4.9	3.1	4.9
Parathion	11.3	2.3	0.8	5.4	0.8	4.1

[a] Precision achieved for 2 mL vials from five repetitive injections of 1.4 mL samples (n = 5). The concentration was 300 μg/L for each compound. A 30 min absorption time was used.
[b] Precision achieved for 16 mL vials from five repetitive injections of 12 mL samples (n = 5). The concentration was 300 μg/L for each compound. A 30 min absorption time was used.
From Ref. 14. Reprinted from Journal of Chromatography A, 1997 with kind permission of Elsevier Science – NL, Sara Burgerhartstraat 25, 1055 KV Amsterdam, The Netherlands.

This absorption process for four s-triazines and parathion when using a flow-through cell shows efficiency similar to other agitation techniques. The samples were pumped at a flow rate of 10 mL/min, achieving a high linear flow of the aqueous sample inside the Teflon tubing where the SPME fiber is positioned during the absorption. The precision achieved for this technique is satisfactory (RSD < 8%, n = 5).

On-line Screening System

A prototype analytical system was developed that allows the quasi-continuous monitoring of organic contaminants in surface water; furthermore, it could also be applicable to the analysis of sewage water [31]. It consists of a flow-through cell and an automated SPME unit, coupled in-line to a gas chromatograph. This system combines the advantages of SPME as a simple, fast, sensitive, and solvent-free sample introduction technique, with the advantages of aqueous sample on-line processing as a less time-consuming, efficient, and continuous technique. Organochlorine pesticides and triazine herbicides were selected to evaluate the system. The flow-through cell was shown to provide a successful way for automated on-line SPME coupled in-line to GC with a precision of approximately 10% RSD for the investigated triazines.

The flowing surface of sewage water was simulated by a spiked, aqueous sample stored in a brown, glass bottle. It is continuously pumped through the cell at 300 mL/min with a peristatic pump. The fiber is dipped at regular intervals into the flowing sample. In the future, a bypass of a river or a sewage effluent will be pumped through the cell, thus allowing the direct and quasi-continuous monitoring of organic compounds at trace levels in surface waters.

A flow-through cell made of glass was constructed by us for the on-line solid-phase microextraction. This cell was mounted onto the commercial auto-sampler modified to take up the cell as shown schematically in Figure 8.3. For this arrangement, nine positions for autosampler vials had to be removed. The glass, flow-through cell forms a half-circle with the absorption position in the middle. At this position, which is arranged exactly at the site of one of the original autosam-pler vials, the cell is sealed by a septum. The fiber, fixed to the SPME-autosampler, dips directly through the seal into the cell. This allows the use of the normal software procedure to also control the entire absorption and desorption steps for this flowing sample arrangement.

For sampling, the fiber is first exposed to the aqueous sample pumped at a constant flow through the cell for a given period. Second, the fiber is withdrawn from the sample and introduced automatically into the GC-injector, where thermal desorption occurs. After desorption, the fiber is withdrawn from the injector and again automatically dipped into the flowing sample. While the next sample is ab-sorbed, the preceding sample is chromatographed. This overlapping of absorption and chromatography reduces the average time for overall analysis of each sample.

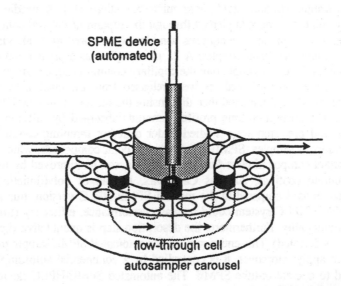

Figure 8.3 Detailed view of the flow-through cell mounted on an autosampler carousel. (From Ref. 31). Reprinted from Journal of Chromatography A, 1996 with kind permission of Elsevier Science – NL, Sara Burgerhartstraat 25, 1055 KV Amsterdam, The Netherlands.

Extensive testing of the developed system shows a good performance, particularly with respect to reproducibility and ruggedness. Using this approach, 150 extractions were performed with a single fiber showing neither a significant deterioration of its performance, nor any mechanical damage. This ruggedness is of particular importance for an on-site operation of the system (e.g., at an effluent of a waste water plant). In this case, the automatic system may be run without operator intervention. Remote control is possible (e.g. transmitting the data via modem to a laboratory). Furthermore, the daily performance of the system can be controlled by continuously adding a marker compound as the internal standard that gives information on the overall performance and quality control of the entire system. The repeatability of this automatic enrichment and analysis system was tested using ten successive extractions, achieving a reproducibility of 4–13% RSD for triazine herbicides.

SPME–HPLC

An automated, SPME-HPLC system was recently developed [15]. Polar, thermally-labile analytes, such as phenylurea pesticides, were selected for microextraction directly from an aqueous sample. A piece of ordinary capillary GC column with its coating (Omegawax 250) was used to absorb the analytes from the aqueous sample (in-tube solid-phase microextraction) [15]. A needle hosts the capillary column when it is pierced through the septum of the vial containing the spiked aqueous sample. The aqueous samples were stored in 2 mL vials on the tray of a commercial autosampler. A 25 μL sample was aspirated and dispensed several times from the sample into the capillary column using a syringe. After the extraction, the absorbed analytes were released from the coating by aspirating methanol into the column, and then dispensing the methanol into the HPLC injector loop. The absorption-time profiles, amounts absorbed by different coatings, linearity, and precision were studied under different sampling conditions using spiked, aqueous samples. SPME selectivity for polar compounds, which represent an important compound class for water analysis, can be improved by using more polar column coatings, such as Carbowax, instead of poly(dimethylsiloxane) coated columns. Compared to the manual SPME-HPLC version, this automated micro SPME/HPLC system could increase performance, efficiency (throughput), and reproducibility. Furthermore, the desorption step is quantitative (i.e., no carryover was detected). This entire method for automated SPME sample preparation is easy to apply; moreover, it is controlled by a commercial autosampler that is modified to operate in-tube SPME. The automated SPME-HPLC device obtains RSDs for all investigated compounds below 6%.

Direct extraction from the aqueous matrix, with the SPME fiber directly exposed to the sample, is typically used for semi-volatile compounds, including polar pesticides such as triazines. The SPME fiber methods developed so far have had limited effectiveness for thermally labile compounds (e.g., phenylurea and carbamate pesticides). These compounds could be extracted from an aqueous matrix only with a very polar coating like Carbowax. Besides, the high polarity of the SPME fiber has to be stable against solvents used in HPLC, such as methanol or acetonitrile. The capacity of the commercially available fibers (total fiber volume times partition coefficient of the stationary phase) is low. Thus, the SPME-HPLC method is characterized by a low sensitivity and low ppb range when using UV detection. Sensitivity can be increased instrumentally by using hyphenated techniques, in particular LC-MS.

The newly-developed, automated SPME-HPLC method uses a flow-through process that is expected to reduce the total extraction time per sample and increase the precision of the entire method. The open tubular column that is used for the direct absorption of the target analytes from the aqueous sample is enclosed in a needle device and can automatically be exposed to a vial containing the sample. The device combines features of earlier developed SPME devices where the inner surface of a syringe is coated with a polymer and microcolumn LC. The absorption equilibrium will be achieved faster because the extraction process involves agitation by sample flow in and out of a column.

Autosampler Operation and Interface Design

The autosampler software is programmed by user-defined programs that control the SPME absorption and desorption. For desorption, pure methanol from a second vial is flushed through the SPME unit and directly transferred into the injection loop. The HPLC injection loop is built of 56 cm long PEEK tubing (300 μm ID) that has a total volume of 40 μL. The instrumental set-up is illustrated in Figure 8.4.

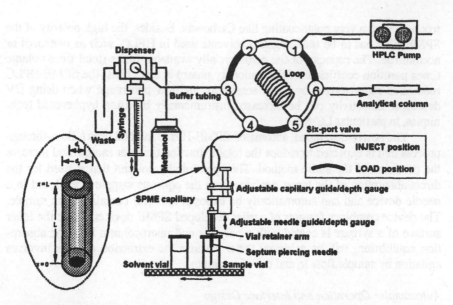

Figure 8.4 Instrumental set-up of the new on-line SPME/HPLC interface based on an in-tube SPME capillary technique. A piece of GC column (in-tube SPME) hosts in the position of the former needle capillary. The aqueous sample is frequently aspirated from the sample vial through the GC column and dispensed back to the vial (INJECT position) by movement of the syringe. After the extraction step, the six-port valve is switched to the LOAD position for the desorption of the analytes from the in-tube SPME by flushing methanol from another vial through the SPME capillary. The volume is transferred to the loop. After switching the Valco valve to the INJECT position an isocratic separation using a mixture of 60/40 aceto-nitrile/water is performed. (From Ref. 15) Reprinted with permission from Analytical Chemistry. Copyright 1995 American Chemical Society.

The 2 mL vials are filled with 1.4 mL of aqueous sample for the absorption of the compounds. The first step in the method is to rinse the GC capillary column with methanol, which it will still contain before the first absorption step. A 25 μL sample volume (total volume of the syringe used in this study) is aspirated from the sample vial at a flow rate 63 μL/min. Then, the same sample volume is dispensed back into the vial. Usually, these two steps are repeated ten times. After the absorption step, the six-port valve is switched to the load position. Then, 38 μL of methanol is aspirated from a solvent vial and transferred to the injection loop for the desorption of the extracted analytes from the capillary coating. The six-port valve is switched to the inject position, and a trigger is sent to the PC for starting the data acquisition. The sample is then transferred from the loop to the analytical column by the isocratic eluent mixture. During the analysis of the first sample, a subsequent sample can be extracted.

In our final experimental design, the sample is aspirated and dispensed instead of a one-way flow. The contamination of the buffer tubing makes the flow-through approach less efficient compared to the repeated aspirate/dispense mode.

The increase in sample flow rate is a significant factor that determines the equilibration time. However, the increase of this parameter is limited by practical factors. Furthermore, the total volume that can be aspirated in one step is limited by the syringe volume. A continuous, one-way aspiration system shows no significant advantages and further improvement of the extraction efficiency. The aspiration concept, however, could improve the automated, in-tube SPME handling substantially by obtaining a higher precision.

The absorbed compounds are first desorbed and then transferred into an HPLC loop. The injection into the analytical column is similar to the injection of methanolic standards. The separation of the desorption step from the injection to HPLC step is a big advantage compared to the manual SPME-HPLC interface. Using this procedure, desorption can be accomplished using a different solvent mixture that shows a higher eluting power, compared to the initial solvent composition of the HPLC eluent.

Table 8.3 Precision and limit of detection (LOD) of the in-tube SPME/HPLC-UV system for six phenylureas.

Compound	M_n[a]	K_{OW}	RT [min]	$w_{1/2}$ [s]	A-wax[b,c]	A-wax[b,d]	SPB-5[b,e]	LOD [µg/L][f]
					Precision (RSD [%])			
Monuron	198	87	3.69	10.4	5.4	3.1	8.3	3.3
Fluometuron	232	263	4.28	9.6	5.6	3.7	4.5	3.3
Diuron	232	479	4.70	8.7	2.6	3.1	3.2	2.7
Siduron	232	1230	5.74	10.6	2.7	2.6	4.2	3.8
Linuron	248	1585	6.72	12.7	2.1	2.8	1.9	2.8
Neburon	274	6310	10.08	17.9	1.6	1.9	2.6	4.1

[a] M_n = nominal mass.
[b] Ten repetitive injections (n = 10).
[c] The concentration in the aqueous sample was 10,000 µg/L for each compound (for an Omegawax 250 capillary).
[d] The concentration in the aqueous sample was 1,000 µg/L for each compound (for an Omegawax 250 capillary).
[e] The concentration in the aqueous sample was 1,000 µg/L for each compound (for a SPB-5 capillary).
[f] Signal-to-noise ratio of S/N = 3. The concentration in the aqueous sample was 10 µg/L for calculating the LOD values for an Omegawax 250 capillary.

(From Ref. 15.) Reprinted with permission from Analytical Chemistry. Copyright 1997 American Chemical Society.

In general, the precision obtained using an Omegawax 250 capillary column was < 6% RSD (n = 10). Table 8.3 shows the precision for all compounds investigated under different conditions. The method was linear over at least three orders of magnitude. The coefficient of correlation achieved was better than 0.99. Using ten aspirate/dispense steps, the limit of detection (LOD) for all compounds was below 5 µg/L. This limit can be optimized by increasing sampling time to achieve equilibrium conditions. Furthermore, the LOD can be lowered by using narrow bore HPLC columns achieving smaller peaks or more sensitive detection methods such as mass spectrometry (MS) in the selected-ion monitoring (SIM) mode. When using hyphenated detection techniques (e.g. LC-MS) the LOD is expected to be below the 0.1 µg/L level.

The SPME absorption and desorption steps are fully automated with the HPLC autosampler. Once programmed, an operator can fill the autosampler tray with the samples and all subsequent steps are controlled by the software. Thus, the newly developed instrument described [15] is expected to show good potential for routine analyses. Furthermore, the selectivity for target analytes, such as drugs or chiral target compounds, could be increased by using very polar or chiral coating materials. For more specific analyses of metabolites and unknown compounds, more selective, hyphenated detection systems (e.g., LC-ESI/MS or LC-API/MS) can be evaluated in the future. A modified, continuous-flow set-up could be helpful for on-line monitoring of target compounds for process control using a modified in-tube SPME device. The automated, micro SPME-HPLC system can be applied for routine analysis of polar, thermally-labile compounds and easy coupling to µ-LC and capillary electrophoresis (CE).

APPLICATIONS BASED ON NEW SPME METHODS

Air Analysis by Integrated Sampling

Integrating sampling is possible with a simple SPME device in addition to the standard analyte concentration measurement at a well defined place in space and time. This type of sampling is particularly important in field studies, when changes of the analyte concentration in time and space has to be taken into account [25]. For integrated sampling, the SPME fiber is not directly exposed to the matrix. Instead, it is kept inside a protective tubing (needle) without any flow of the sample through it. The extraction occurs though a static layer of gas present in the needle. The system can be designed either as the in-tube SPME discussed for the SPME/HPLC system where the extracting phase is coating the interior of a tubing, or as a fiber coated with the stationary phase that is placed inside the protective needle. The latter device is the classical, manual SPME fiber device, except it is used without exposing the fiber during the sampling. The position of the fiber inside the protecting sheath is very important because it has to be placed accurately at the same position that represents the same volume of protecting gas. The mechanism of analyte transport into the extracting phase in such a set-up is determined only by the diffusion through the gaseous phase. Thus, the response is proportional to the integral of the analyte concentration over the time and space (when the needle is moved through space) [21]. The process is characterized by a linear concentration profile established in the tubing between the small needle opening with the area, A, and the position of the extracting phase that is located at the position Z from the opening. The amount of analyte extracted, dn, during the time interval, dt, can be derived from Fick's law of diffusion [32] and can be expressed by:

$$dn = AD_g \frac{dc}{dz}dt = AD_g \frac{\Delta C(t)}{Z}dt \qquad 8.2$$

where $\Delta C(t)/Z$ is a value of the gradient established in the needle between the needle opening and the position of the extracting phase, Z; $\Delta C(t) = C(t) - C_Z$, where $C(t)$ is a time dependent concentration of the analyte in the sample in the vicinity of the needle opening, and C_Z is close to zero for a high coating/gas distribution constant capacity, then: $\Delta C(t) = C(t)$. D_g is the diffusion coefficient of the analyte in the gaseous phase. The concentration of analyte at the coating position in the needle, C_Z, will increase with integration time, but it will be kept low compared to the sample concentration. Thus, the accumulated amount over time can be calculated as:

$$n = D_g \frac{A}{Z}\int C(t)dt \qquad 8.3$$

Equation 8.3 shows that the extracted amount of analyte is proportional to the integral of the sample concentration over time. This equation is valid for the assumption that the amount absorbed onto the extraction sorbent is a small fraction (below the RSD of the measurement, typically 5%) of equilibrium amount with respect to the lowest concentration in the sample. By varying the position of the sorbent in the tubing, the integration time can be extended. The sensitivity of the system can be improved by increasing the stationary phase volume or by using more selective coating materials with larger coating/gas distribution constants.

The integrated, in-tube SPME sampling device was applied to field sampling [25] for determining concentrations of styrene in indoor air at an industry due to the application of large quantities of vinyl ester resin. The 30 minute integrated sampling using a 100 μm PDMS fiber shows good agreement with additional determinations based on direct SPME grab sampling and 30 minute charcoal tube sampling results [25]. The concentrations determined were in the 100 μg/L range. SPME for fast grab sampling and integrated sampling can be an alternative method to the time-consuming active charcoal sampling that needs (toxic) solvents for the desorption process.

Hot Water Extraction for the Analysis of Semivolatiles from Soil

Efficient extraction of semivolatile organic compounds from soil samples can be obtained using supercritical (high temperature) water to extract the target analytes that are later extracted from water by SPME [33,34]. Two techniques are reported in the literature for determining PAHs in soils and urban air particulate. Achieving recoveries of 60 to 140%, the method uses supercritical water extraction at 250°C for 15–60 min [33]. The SPME fiber is exposed to the cooled water sample to trap the target analytes. The second method combines both steps, thus the fiber is placed inside the cell during the dynamic hot water extraction [34]. The apparatus used for this application is shown in Figure 8.5. The method was validated using NIST standard material for PAHs in urban air particulate [34].

Fast Gas Chromatography

Solid-phase microextraction has been successfully used as the sample introduction technique for fast-gas chromatography. A modified injector is used that allows fast heating by applying capacity discharge to a heating wire [35]. Furthermore, flash heating is achieved by passing an electric current directly through a wire used in place of the fiber or by using a hollow fiber equipped with an internal micro heater. Using this system, BTEX separation is achieved in less than 9 seconds, and volatile organic compounds from EPA method 624 are separated within 150 seconds. The SPME-Fast GC set-up was also installed on a portable, field GC instrument, which is discussed in the next section. Using a temperature-programmed separation, Purgeables A and B could be separated within 1.5 min.

Fast GC demonstrates a high efficiency when coupled with an extraction technique that can be performed within the same time frame. On the other hand, a Fast GC system is not necessary when the long extraction time of the sample preparation method determines the total analysis time. SPME combines the unique feature of fast sample preparation and easy injection into the GC system. Thus, the coupling of both techniques shows a significant increase on the sample throughput for the analysis. Furthermore, the sample handling is very easy and rugged, which makes the technique amenable for field sampling. Typical sample turnaround is 5 minutes with good method precision. The interface for SPME/Fast-GC is available from SRI Instruments.

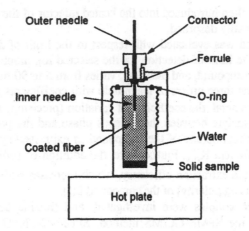

Figure 8.5 Schematic diagram of a high pressure extraction cell for static high temperature water extraction with simultaneous SPME. (From Ref. 34).

Field Analysis

A commercially-available, field-portable gas chromatograph has been adapted to enable the use of SPME as the sample preparation and introduction technique for fast GC in the field [16]. Fast GC separations of BTEX compounds in less than 15 seconds were reported. Thus, four BTEX compounds, which were extracted by SPME on a PDMS/DVB fiber, were determined using a modified photoionization detector (PID). The analysis of trichloroethene in soil samples was used to evaluate the instrument in the field. The SPME method was used for 500 samples analyzed within 5 days, showing a high ruggedness of the method and instrumentation during on-site monitoring. Taking into account the versatility of SPME, this technique enables preliminary screening of a wide range of compounds and/or samples (aqueous, gaseous, solid) directly in the field. These results are required to determine whether a suspect sample has to be drawn and further analyzed.

Multi–Residue Methods for Pesticides

Multi-residue methods have been developed for a large number of pesticides covering different compound classes [30,36]. One method, based on the manual SPME device, was developed to determine nitrogen- and phosphorous-containing pesticides (amines, anilides, phosphorothioates, and triazines) by coupling SPME to GC/MS [30]. An 85 μm poly(acrylate) fiber is used to sample directly from the aqueous sample, then introduced into the heated injector of the GC/MS where the analytes are thermally desorbed.

The method was evaluated with respect to the limit of detection, linearity, and precision. The limit of detection for the selected ion monitoring (SIM) mode depends on the compound, and therefore, varies from 5 to 90 ng/L. The method is linear over at least three orders of magnitude with coefficients of correlation usually ≥ 0.996. In general, the coefficient of variation (precision) is $< 10\%$. The partitioning of the analyte between the aqueous phase and the polymeric phase depends on the hydrophobicity of the compound as expressed by the octanol-water partitioning coefficient K_{ow}. Furthermore, the addition of sodium chloride has a strong affect on the extraction efficiency, which increases with decreasing hydrophobicity (increasing polarity) of the compound [22].

Real water samples were investigated, and triazine herbicides could be identified using the SPME-GC/MS method. SPME-GC/NPD was first used to identify and quantify the triazines: atrazine, simazine, and terbuthylazine, in water samples from sewage water plants. For such a complex matrix, GC/NPD is not sufficiently selective for an unambiguous identification at low levels (< 1 ppb) of pesticides. Consequently, selectivity is enhanced by using SPME-GC/MS in the SIM mode with three characteristic ions for each pesticide. This method unequivocally identifies and quantifies low levels of pesticides in environmental samples. Moreover, the results were verified by an alternative method, on-line solid-phase extraction (SPE) coupled to LC-MS, which indicated a good correlation to the SPME results.

The use of liquid or solid polymers, such as PDMS or poly(acrylate) coatings, is based on absorption. A high content of matrix compounds that have a high affinity to the fiber coating does not significantly affect the analyte partitioning as was shown in the past [30]. On the other hand, the adsorption phenomena determine the extraction process for the mixed porous polymer coatings, such as DVB or Carbowax. The adsorption is significantly affected by competition phenomena, especially when all active sites are occupied. The higher sensitivity obtained from these new fibers might be accomplished by significant competition at high matrix concentrations, which results in a lower precision. One must decide which point is more important for the analysis: sensitivity or precision.

Furthermore, SPME was successfully used for determining metolachlor in runoff and tile-drainage water [37]. In addition, this method was compared to solid-phase extraction and immunoassay analyses of metolachlor [38], and a good correlation of the three methods was found in this interlaboratory study. Therefore, SPME and ELISA have a high potential for determining metolachlor in natural waters [38].

Applying SPME for analyzing semi-volatile compounds in water was verified by an interlaboratory study on pesticide analysis [39]. The test was done at low ppb levels with participants from eleven laboratories in Europe and North America. The test results proved that SPME is an accurate and fast method of sample preparation and analysis [40].

Liquid Chromatography – Mass Spectrometry (LC/MS)

The first combination of solid-phase microextraction and HPLC-electrospray ionization-mass spectrometry (ESI-MS) was applied for determining polar, water-soluble components from sludge and sediments [41]. The extraction, desorption, and detection conditions were examined for selected carboxylic acids, phthalates, and surfactants using Carbowax-coated SPME fibers. Maximum extraction yields of the target compounds were obtained by extracting between 1 to 15 hours using magnetic stirring (phthalates and surfactants) and desorbing for two minutes with a methanol/ethanol mixture of 80:20 (v/v) [41]. The main components of the analyzed sludge and sediment samples were identified as phthalates, fatty acids, non-ionic surfactants, chlorinated phenols, and carbohydrate derivatives. Additional information for compound identification was obtained at different spray potentials for the mass spectrometric ionization. Switching from positive to negative ionization mode changes the detection selectivity and sensitivity; therefore, substances of different natures could be detected.

A water slurry (4 mL) was prepared from 10 mg of each of the dried and homogenized solid samples. The slurries were saturated with sodium chloride and adjusted to pH 2 with 25% hydrogen chloride solution to force the dissociation equilibrium of the polar compounds to the non-dissociated species. For desorption, 100 μL of a methanol/ethanol (80:20) mixture was injected into the injection port before the fiber was placed and fixed in the desorption chamber. The use of atmospheric pressure ionization mass spectrometry avoids pre-derivatization and guarantees a high sensitivity, especially for ionic, thermal-labile, and polar compounds. The electrospray source was operated in positive as well as negative ionization mode to utilize the different selectivity of both options. In the positive mode, (ESI^+), mainly cations are recorded or ions that dispose of sufficient proton affinities to produce the $[M+H]^+$ ions. The positive ESI mode proved to be an appropriate ionization mode for the phthalates (DBP and DEHP) studied. Negative spray conditions abstract protons from the analytes, which leads to $[M-H]^-$ ions. This process is dominant for acidic compounds like fatty acids, diacids, and alkyl-polyglycoside.

The $[M+H]^+$ and $[M-H]^-$ ions were used for target analysis. These ions are of diagnostic value because of their dominance in the mass spectra. Alkyldiacids, like butanediocacid and succinates, adipic acid, and corresponding esters known as degradation products of biological processes (trace m/z 133, 147) were recorded. Alkylbenzenesulphonates (m/z 339) and alkylcarboxylic ions with m/z 227, 255, and 425 were identified. These compounds are released typically from soap products and fat metabolism. The chromatogram traces of the corresponding SPME-ESI⁻ analysis show additional sediment constituents identified as surfactant and carboxylic compounds. In addition, trichlorophenol was identified in the sediment.

The method combination of SPME/LC-MS proved to be a very powerful instrumentation for determining water-soluble compounds if current methods like GC/MS or HPLC-UV failed, or if special efforts are necessary for their detection (e.g., derivatization). A 10 mg sample leads to identifying components like trichlorophenol, phthalate derivatives, alkylsulphonate and polyethylether surfactants, fatty acids, diacid esters and carbohydrate derivatives. Different ionization options of electrospray mass spectrometry (ESI⁺ and ESI⁻) were used for substance identification. Carboxylic acid esters, as well as non-ionic surfactants, are more effectively ionized by a positive spray potential. The negative spray mode was successful for the detection of chlorophenols, anionic surfactants, and some carbohydrates. Carboxylic acids could be ionized with both methods.

Speciation of Inorganic Compounds

Speciation of some inorganic compounds can be achieved by solid-phase microextraction. Tetraethyllead (TEL) and ionic lead in water are determined in the lower ppt range by using SPME coupled to GC-FID and GC-MS [17]. TEL is extracted from the headspace of a sample, and inorganic lead is first derivatized using sodium tetraethylborate to form TEL. Dimethylmercury was determined by SPME coupled to gas chromatography and microwave-induced plasma atomic emission detection [18]. Natural gas condensate has also been investigated resulting in a limit of detection around 20 μg/L [18]. Using element-specific detectors, such as an atomic emission detector (AED) or inductively-coupled plasma-mass spectrometry (ICP-MS), a variety of different inorganic species can be easily determined. A sensitive and simultaneous determination of organo-mercury, -lead, and -tin compounds was developed using headspace SPME coupled to GC-ICP/MS [19]. The organometallic compounds are derivatized in-situ with sodium tetraethylborate, then extracted on a poly(dimethysiloxane) fiber. The detection limits obtained when using ICP-MS were 0.34 and 2.1 ng/L as tin for monobutyl-, dibutyl-, and tributyltin [19]. The method was validated using PACS-1 reference materials.

PERSPECTIVE FOR FUTURE APPLICATIONS

The new applications discussed here demonstrate the high potential of solid-phase microextraction for fast screening and precise quantification methods for mainly organic target analytes at trace levels. The following major directions for future investigations can be summarized. Direct air sampling and determining indoor air using integrated sampling techniques can be used as alternative methods in industrial hygiene. The coupling of SPME with Fast GC increases sample handling throughput and shows high applicability for field sampling methods. The dimensions of the micro device used for SPME increases the potential for small sample volumes. The small fibers can be exposed directly inside a flower for determining flavor compounds in air. The equilibrium nature of SPME and typically very small amounts extracted are important when in-situ monitoring with little or no interference of the process is required. The non-disturbing nature of SPME opens new application fields where exhaustive extraction techniques, such as SPE, are not amenable (e.g., the study of flowers' fragrance spectra or determining partitioning constants in complex systems without affecting the system [42]). In addition to more elegant derivatization techniques that are already used for SPME, the automated, in-tube SPME/HPLC method increases the potential for polar, thermally labile compounds that can be determined in very small sample volumes. This opens applications for drug analysis where the sample volume is limited and continuous monitoring becomes an important issue, such as intoxication, degradation, and metabolism studies. SPME has been successfully coupled to capillary electrophoresis (CE) analyzing PAHs [43] and barbiturates [44]. In addition, fully automated systems might be helpful for the design of continuous and quasi-continuous analysis apparatus that can be operated in remote field situations or on-site, for process control or monitoring of river and surface water systems. Consequently, the data base of these studies and its reliability will be increased. Fast GC can be easily performed in portable, micro GCs when using SPME for the sample preparation. New coating materials, such as metal coatings and porous polymers, will increase the number of specific applications. Furthermore, the technique is suitable for determining physical constants, such as distribution constants with minimum or no interference of the target system. This approach was shown for determining distribution constants of chemicals in water in the presence of humic dissolved organic matter (HOM) [42]; moreover, it could be extended to even very complex systems, such as investigating drug distribution in body fluids. The SPME fiber technique can be used for direct coupling to MS and ICP/MS [19], gaining maximum selectivity for determining organic and inorganic target compounds. The direct coupling without any column separation technique increases the sample throughput significantly and increases sensitivity.

The unique, solvent-free, and easy sample preparation method of SPME has been successfully applied to many organic target analytes in environmental, bioanalytical, and industrial hygiene studies. Further automation and miniaturization of the entire tool will keep the advantages of this extraction device.

ACKNOWLEDGMENT

The authors would like to thank Supelco Canada Inc., Varian Canada Inc., and the National Sciences and Engineering Research Council of Canada (NSERC) for financial support for many parts of their research. The authors wish to gratefully acknowledge instrument loans from LC Packings (Amsterdam, The Netherlands) and SRI Instruments Inc. (Las Vegas, NV).

REFERENCES

1. J Pawliszyn. Solid-Phase Microextraction: Theory and Practice, New York: Wiley-VCH, Inc., 1997.
2. RP Belardi and J Pawliszyn. The application of chemically modified fused silica fibers in the extraction of organics from water matrix samples and their rapid transfer to capillary columns. Water Pollut Res J Can 24:179-191,1989.
3. CL Arthur and J Pawliszyn. Solid phase microextraction with thermal desorption using fused silica optical fibers. Anal Chem 62:2145-2148,1990.
4. CL Arthur, LM Killam, KD Buchholz, J Pawliszyn, and JR Berg. Automation and optimization of solid-phase microextraction. Anal Chem 64:1960-1966, 1992.
5. D Louch, S Motlagh, and J Pawliszyn. Dynamics of organic compound extraction from water using liquid-coated fused silica fibers. J Anal Chem 64:1187-1199, 1992.
6. M Chai, CL Arthur, J Pawliszyn, RP Belardi, and KF Pratt. Determination of volatile chlorinated hydrocarbons in air and water with solid phase microextraction. Analyst 118:1501-1505, 1993.
7. Z Zhang and J Pawliszyn. Headspace solid phase microextraction. Anal Chem 65:1843-1852, 1993.
8. KD Buchholz and J Pawliszyn. Determination of phenols by solid phase microextraction and gas chromatographic analysis. Environ Sci Technol 27:2844-2848,1993.
9. KD Buchholz and J Pawliszyn. Optimization of solid-phase microextraction (SPME) conditions for phenol analysis. Anal Chem 66:160-167, 1994.
10. J Chen and JB Pawliszyn. Solid phase microextraction coupled to high performance liquid chromatography. Anal Chem 67:2530-2533,1995.
11. CL Arthur, LM Killam, S Motlagh, M Lim, DW Potter, and J Pawliszyn. Analysis of substituted benzene compounds in ground-water using SPME. Environ Sci Technol 26:979-983, 1992.
12. P Popp, K Kalbitz, and GJ Oppermann. Application of solid-phase microextraction and gas chromatography with electron-capture and mass spectrometric detection for the determination of hexachlorocyclohexanes in soil solutions. Chromatogr A 687:133-140, 1994.
13. JR Berg. Practical Use Of Automated Solid Phase Microextraction. Am Lab 1993, 25, 18-24.
14. R Eisert and J Pawliszyn. Design of automated SPME for trace analysis of organic compounds in aqueous samples. J Chromatogr A 776:293-303, 1997.
15. R Eisert and J Pawliszyn. Automated in-tube solid phase microextraction coupled to high-performance liquid chromatography. J Anal Chem 69:3140-3147, 1997.
16. T Górecki and J Pawliszyn. Field-portable solid - phase microextraction /fast GC system for trace analysis. J Field Anal Chem Technol, 1(5): 277-284, 1997.

17. T Górecki and J Pawliszyn. Determination of tetraethyllead and inorganic lead in water by solid phase microextraction/gas chromatography. Anal Chem 68:3008-3014, 1996.
18. JP Snell, W Frech, and Y Thomassen. Performance improvements in the determination of mercury species in natural gas condensate using an on-line amalgamation trap or solid-phase microextraction with capillary gas chromatography-microwave-induced plasma atomic emission spectrometry. Analyst 121:1055-1060, 1996.
19. L Moens, TD Smaele, R Dams, P van den Broeck, and P Sandra. Simultaneous determination of organomercury, -lead, and -tin compounds with headspace solid phase microextraction capillary gas chromatography combined with inductively coupled plasma mass spectrometry. Anal Chem 69:1604-1611, 1997.
20. AD Young. Boundary Layers, Oxford: BSP Professional books, 1989.
21. M Chai and J Pawliszyn. Analysis of environmental air samples by solid phase microextraction and gas chromatography/ion trap mass spectrometer. Environ Sci Technol 29:693-701, 1995.
22. R Eisert and K Levsen. Determination of organophosphorus, trizazine and 2,6-dinitroanaline pesticides in aqueous samples via solid-phase microextraction (SPME) and gas chromatography with nitrogen-phosphorus detection. Fresenius' J Anal Chem 351:555-562, 1995.
23. JJ Langenfeld, SB Hawthorne, and DJ Miller. Quantitative analysis of fuel-related hydrocarbons in surface water and wastewater samples by solid-phase microextraction. Anal Chem 68:144-155, 1996.
24. PA Martos, A Saraullo, and J Pawliszyn. Estimation of air/coating distribution coefficients for solid phase microextraction using retention indexes from linear temperature-programmed capillary gas chromatography. Application to the sampling and analysis of total petroleum hydrocarbons in air. Anal Chem 69:402-408, 1997.
25. PA Martos and J Pawliszyn. Calibration of solid phase microextraction for air analyses based on physical chemical properties of the coating. Anal Chem 69:206-215, 1997.
26. A Saraullo, PA Martos, and J Pawliszyn. Water analysis by solid phase microextraction based on physical chemical properties of the coating. Anal Chem 69:1992-1998, 1997.
27. Z Zhang, J Poerschmann, and J Pawliszyn. Direct solid phase microextraction of complex aqueous samples with hollow fiber membrane protection. Anal Commun 33:219-221, 1996.
28. S Motlagh and J Pawliszyn. On-line of flowing samples using solid phase microextraction-gas chromatography. Anal Chim Acta 284:265-273, 1993.
29. AA Boyd-Boland and JB Pawliszyn. Solid-phase microextraction coupled with high-performance liquid chromatography for the determination of alkylphenol ethoxylate surfactants in water. Anal Chem 68:1521-1529, 1996.

30. R Eisert and KJ Levsen. Determination of pesticides in aqueous samples by solid-phase microextraction in-line coupled to gas chromatography-mass spectrometry. Am Soc Mass Spectrom 6: 1119-1130, 1995.
31. R Eisert and KJ Levsen. Development of a prototype system for quasi-continuous analysis of organic contaminants in surface or sewage water based on in-line coupling of solid-phase microextraction to gas chromatography. Chromatogr A, 737:59-65, 1996.
32. Crank, J. Mathematics and Diffusion, Oxford: Clarendon Press, 1989, p. 14.
33. KJ Hageman, L Mazeas, CB Grabanski, DJ Miller, and SB Hawthorne. Coupled subcritical water extraction with solid-phase microextraction for determining semivolatile organics in environmental solids. Anal Chem 68:3892-3898,1996.
34. H Daimon and J Pawliszyn. High temperature water extraction combined with solid phase microextraction. Anal Commun 33:421-424, 1996.
35. T Górecki and J Pawliszyn. Sample introduction approaches for solid phase microextraction/rapid GC. Anal Chem 67:3265-3274, 1995.
36. AA Boyd-Boland, S Magdic, and J Pawliszyn. Simultaneous determination of 60 pesticides in water using solid phase microextraction and gas chromatography-mass spectrometry. Analyst 121:929-938,1996.
37. KN Graham, LP Sarna, GRB Webster, and JD Gaynor. Solid-phase microextraction of the herbicide metolachlor in runoff and tile-drainage water samples. J Chromatogr A 725:129-136, 1996.
38. JD Gaynor, DA Cancilla, GRB Webster, LP Sarna, KN Graham, HJF Ng, CS Tan, CF Drury, and TJ Welacky. Comparative solid phase extraction, solid phase microextraction, and immunoassay analyses of Metolachlor in surface runoff and tile drainage. Agric Food Chem 44:2736-2741, 1996.
39. T Górecki, R. Mindrup, and J Pawliszyn. Pesticides by solid-phase microextraction. Results of a round robin test. Analyst 121:1381-1386, 1996.
40. R Eisert, T Górecki, and J Pawliszyn. Pesticide analysis by solid-phase microextraction. Am Environ Lab 4: 20-21, 1997.
41. M Moder, P Popp, and J Pawliszyn. Characterization of water-soluble components of slurries using solid-phase microextraction coupled to liquid chromatography-mass spectrometry. J Microcol Sep, 10(2):225-234, 1998.
42. J Poerschmann, Z Zhang, FD Kopinke, and J Pawliszyn. Solid phase microextraction for determining the distribution of chemicals in aqueous matrixes. Anal Chem 69:597-600, 1997.
43. AL Nguyen and HT Luong. Separation and determination of polycyclic aromatic hydrocarbons by solid phase microextraction/cyclodextrin-modified capillary electrophoresis. Anal Chem 69:1726-1731,1997.
44. S Li and G Weber. Determination of barbiturates by solid phase microextraction and capillary electrophoresis. Anal Chem 69:1217-1222, 1997.

Appendix

New SPME developments and applications using this innovative technique are continually being developed and published. Several excellent sources for further information are listed here.

http://sciborg.uwaterloo.ca/chemistry/pawliszyn/

This University of Waterloo website contains information on new developments from the research group lead by Prof. Janusz Pawliszyn, the inventor of SPME.

http://www.cm.utexas.edu/~brodbelt/spme_refs.html

This site has a complete listing of published SPME articles. It is part of the Brodbelt Research Group site at the University of Texas in Austin, who has used SPME extensively in their research.

http://www.varianinc.com/csb/apps.html

Varian, Inc. has an extensive list of SPME application notes focusing on automating the technique. This site also includes information on other GC, GC/MS, and HPLC applications.

https://www.sigma-aldrich.com/SAWS.nsf/Pages/Supelco?EditDocument

This Supelco website provides a comprehensive list of SPME applications including frequently asked questions to improve system performance.

Index

Absorption (*see also* Extraction,
  Sampling), optimizing time,
  45
Accelerants:
  in human skin, 204-205
  in water, 204
Accelerated solvent extraction, 133
Acenaphthene:
  in sand, 9, 12
  in water, 24, 41-42
Acenaphthylene:
  absorbed on various PDMS film
    thicknesses, 69-70
  in sand, 9, 12
  in water, 24
Acetic acid:
  influence of pH, 181-182
  in wine, 83, 85-86
Acetone:
  analysis of:
    comparing SPME fibers, 115-
    116
    as a residual solvent, 115-124
  in air, 95-96
  in human breath, 211
  in water, 76-77, 81
    large versus small vials, 44-45
Acetonitrile:

analysis of :
  in air, 95-96
  as a residual solvent, 115-124
  in water, 81, 90-91
Acid herbicides, analysis of, 103
Acids, analysis of, 99, 179-181, 187
Air monitoring (*see also* Sampling,
    field):
  analysis of, 155-159
  traditional methods, 135
  using integrated sampling, 232-233
Alcohols, analysis of, 99, 179-181
Aldehydes, analysis of, 99, 179-181
Alkanes:
  analysis of, 82-83
  in sand, 9, 11
  in water, 73-77, 86-90
Alkylphenol ethoxylates in water,
  173
American Society for Testing and
  Materials, 132
Ametryn, in water, 18, 20-21
Amines:
  analysis of, 90-91, 99
  aromatic, analysis of, 99
Amitryptiline hydrochloride, analysis
    for OVIs, 118-119
Amphetamines:

245

T - #0181 - 101024 - C0 - 229/152/15 [17] - CB - 9780824770587 - Gloss Lamination